Communications in Computer and Information Science

2599

Series Editors

Gang Li, *School of Information Technology, Deakin University, Burwood, VIC, Australia*

Joaquim Filipe, *Polytechnic Institute of Setúbal, Setúbal, Portugal*

Zhiwei Xu, *Chinese Academy of Sciences, Beijing, China*

Rationale

The CCIS series is devoted to the publication of proceedings of computer science conferences. Its aim is to efficiently disseminate original research results in informatics in printed and electronic form. While the focus is on publication of peer-reviewed full papers presenting mature work, inclusion of reviewed short papers reporting on work in progress is welcome, too. Besides globally relevant meetings with internationally representative program committees guaranteeing a strict peer-reviewing and paper selection process, conferences run by societies or of high regional or national relevance are also considered for publication.

Topics

The topical scope of CCIS spans the entire spectrum of informatics ranging from foundational topics in the theory of computing to information and communications science and technology and a broad variety of interdisciplinary application fields.

Information for Volume Editors and Authors

Publication in CCIS is free of charge. No royalties are paid, however, we offer registered conference participants temporary free access to the online version of the conference proceedings on SpringerLink (http://link.springer.com) by means of an http referrer from the conference website and/or a number of complimentary printed copies, as specified in the official acceptance email of the event.

CCIS proceedings can be published in time for distribution at conferences or as post-proceedings, and delivered in the form of printed books and/or electronically as USBs and/or e-content licenses for accessing proceedings at SpringerLink. Furthermore, CCIS proceedings are included in the CCIS electronic book series hosted in the SpringerLink digital library at http://link.springer.com/bookseries/7899. Conferences publishing in CCIS are allowed to use Online Conference Service (OCS) for managing the whole proceedings lifecycle (from submission and reviewing to preparing for publication) free of charge.

Publication process

The language of publication is exclusively English. Authors publishing in CCIS have to sign the Springer CCIS copyright transfer form, however, they are free to use their material published in CCIS for substantially changed, more elaborate subsequent publications elsewhere. For the preparation of the camera-ready papers/files, authors have to strictly adhere to the Springer CCIS Authors' Instructions and are strongly encouraged to use the CCIS LaTeX style files or templates.

Abstracting/Indexing

CCIS is abstracted/indexed in DBLP, Google Scholar, EI-Compendex, Mathematical Reviews, SCImago, Scopus. CCIS volumes are also submitted for the inclusion in ISI Proceedings.

How to start

To start the evaluation of your proposal for inclusion in the CCIS series, please send an e-mail to ccis@springer.com.

Yetunde Folajimi · Leonidas Deligiannidis ·
Salem Othman · Shawren Singh

Editors

Social Impact of AI: Research, Diversity and Inclusion Frameworks

International Workshop, SIAI 2025
Philadelphia, PA, USA, March 1, 2025
Proceedings

 Springer

Editors
Yetunde Folajimi
Wentworth Institute of Technology
Boston, MA, USA

Leonidas Deligiannidis
Wentworth Institute of Technology
Boston, MA, USA

Salem Othman
Wentworth Institute of Technology
Boston, MA, USA

Shawren Singh
Wentworth Institute of Technology
Boston, MA, USA

ISSN 1865-0929   ISSN 1865-0937 (electronic)
Communications in Computer and Information Science
ISBN 978-3-031-98948-3   ISBN 978-3-031-98949-0 (eBook)
https://doi.org/10.1007/978-3-031-98949-0

This Springer imprint is published by the registered company Springer Nature Switzerland AG
The registered company address is: Gewerbestrasse 11, 6330 Cham, Switzerland

If disposing of this product, please recycle the paper.

Preface

The emergence of Artificial Intelligence (AI) as a transformative technology has reshaped nearly every aspect of modern life—from health and education to commerce, governance, and interpersonal communication. While this acceleration has brought about remarkable innovations, it has also surfaced critical questions about ethics, accountability, transparency, and inclusion. Who is AI serving? Whose voices are being amplified—or silenced? What frameworks can ensure that AI systems reflect and respect the diversity of human societies?

The Social Impact of AI: Research, Diversity, and Inclusion Frameworks (SIAI-ReDI) workshop series was created to grapple with these very questions. Since its inception, SIAI-ReDI has fostered a collaborative, interdisciplinary community of scholars, developers, educators, policymakers, and advocates. Now in its fourth edition, SIAI-ReDI 2025 was held on March 1, 2025, at the Pennsylvania Convention Center in Philadelphia, as part of the 39th AAAI Conference on Artificial Intelligence. This full-day event was organized under the conference's Diversity and Inclusion track and brought together a global audience with diverse disciplinary backgrounds and lived experiences.

SIAI-ReDI 2025 centered on the development and assessment of inclusive AI frameworks that prioritize ethical design, equitable deployment, and cultural context. The workshop's call for papers invited submissions across a range of urgent topics, including:

- Algorithmic fairness and transparency
- Inclusive AI education and workforce strategies
- AI in marginalized and underrepresented communities
- Intersectionality, gender, and accessibility in AI
- Cross-cultural AI governance and regulation
- Public trust, participatory AI, and responsible design

These themes reflect the growing need to embed justice and equity at every stage of AI development—from datasets and models to products and policies.

From a competitive pool of 26 submissions, the program committee accepted 8 full papers, 4 short papers, and one poster, following a single-blind peer review process. The review process was conducted by a global panel of experts, each paper receiving at least three reviews. The accepted papers represent a diversity of methodologies, case studies, and theoretical frameworks, collectively advancing our understanding of how AI intersects with societal structures and human rights.

A major highlight of the workshop was the keynote address by Padmanabhan (Peter) Santhanam, Program Director for Open AI Technology Advocacy at IBM T. J. Watson Research Center. In his talk titled *"An Open Ecosystem to Support AI Value Alignment & Human Feedback"*, Dr. Santhanam addressed the persistent challenge of bias in large-scale foundation models. He proposed practical solutions centered on open data provenance catalogs, dynamic testing and evaluation frameworks, and incremental knowledge feedback loops—highlighting the AI Alliance's initiatives toward trustworthy AI ecosystems. His talk set a visionary tone, emphasizing collaboration and community agency in shaping AI's future.

The workshop also featured a dynamic expert panel chaired by Jason M. Grant of Villanova University, a leading voice in biometrics, computer vision, and diversity advocacy in STEM. The panel brought together academic, industry, and student voices to explore the ethical design and societal implications of AI across global contexts. Panelists included:

- Leon Deligiannidis (Wentworth Institute of Technology), who offered insights on brain-computer interfaces and STEM education.
- Ying Li (Meta), who shared her experience leading AI-driven product innovation across privacy, commerce, and search.
- Mehmet Ergezer (Wentworth Institute of Technology), who emphasized intercultural competence and AI education.
- Alvand Vahediahmar (Drexel University), who discussed adversarial robustness and fairness in AI systems.
- Rohan Kulkarni (Meta), who addressed privacy-enhancing technologies and user-centered security.
- Ibukun Folajimi (Wentworth Institute of Technology), an undergraduate student and rising AI scholar, who shared his experience as a student navigating AI ethics and aspiration.

The panel underscored the value of lived experience, collaboration across sectors, and a multi-generational commitment to inclusive innovation.

We express our heartfelt gratitude to all the authors, reviewers, and attendees whose commitment and contributions made this event a success. Special thanks to our Program Committee and external reviewers for their diligent and thoughtful evaluations, and to the School of Computing and Data Science at Wentworth Institute of Technology for its sponsorship and institutional support.

We also wish to acknowledge the Association for the Advancement of Artificial Intelligence (AAAI) for hosting this workshop as part of its ongoing dedication to equity, access, and responsible AI advancement. The inclusion of SIAI-ReDI in the AAAI conference's Diversity and Inclusion track is a testament to the importance of centering ethical and societal impact in mainstream AI research agendas.

As AI continues to evolve at an unprecedented pace, workshops like SIAI-ReDI are essential for grounding technological progress in social responsibility. The proceedings collected here are not only a reflection of rigorous academic inquiry but also a call to action—a reminder that AI must be built for, with, and by diverse communities. We hope that readers will find this volume both intellectually enriching and practically empowering, and that it serves as a foundation for future research, collaboration, and advocacy.

Yetunde Folajimi
Leonidas Deligiannidis
Salem Othman
Shawren Singh

Organization

Program Committee

Yetunde Folajimi (Chair)	Wentworth Institute of Technology, USA
Leonidas Deligiannidis (Co-chair)	Wentworth Institute of Technology, USA
Salem Othman (Co-chair)	Wentworth Institute of Technology, USA
Shawren Singh (Co-chair)	University of South Africa, South Africa
Ibrahim Joseph	TheLashDepott, USA
Oluwatomisin Arokodare	Georgia Southern University, USA
Taiwo Agbaje	Western Illinois University, USA
Mary Isangediok	GEICO, USA
Tope Amusa	Georgia State University, USA
Chandrasekar Ramachandran	Carnegie Mellon University, USA
Siddharth Raina	Carnegie Mellon University, USA
Shawn Ogunseye	Bentley University, USA
Chika Yinka-Banjo	University of Lagos, Nigeria
Rohan Kulkarni	Meta, USA
Mohammadmahdi Vahediahmar	Drexel University, USA
Sheshananda Reddy Kandula	Adobe, USA
Azmine Toushik Wasi	Shahjalal University of Science and Technology, Bangladesh
Kapil Tomar	IIT Delhi, India
Ashutosh Ahuja	Starbucks Technologies, USA
Zhongdongming Dai	University of California, San Diego, USA
Sravan Chittimalla	Enliven Technology Inc., USA
Mounika Revanuru	Wentworth Institute of Technology, USA
Run Yu	SambaNova Systems, USA
Himaghna Bhattacharjee	Meta Platforms Inc., USA
Anthony Clark	Wentworth Institute of Technology, USA
Mahesh Chayel Mohinder Singh	Meta, USA
Ruchi Sharma	Meta, USA
Ashim Dey	Chittagong University of Engineering and Technology, Bangladesh
Khadijat Ladoja	University of Ibadan, Nigeria

Assessing Adolescents' Perceptions of AI Chatbots for Sexual and Reproductive Health Information in Nigeria

Habeeb Abdulrauf[1,2] and Abdulmalik Adetola Lawal[1,2]

[1] Western Michigan University, Kalamazoo, MI, 49008 USA
[2] University of Nevada, Reno, NV, 89557 USA

Abstract. Adolescents in Nigeria face persistent barriers to accessing sexual and reproductive health (SRH) information, including stigma, privacy concerns, and limited healthcare resources (Ajibade et al., 2022; Nmadu et al., 2020). AI chatbots offer a promising solution by providing confidential, accessible, and immediate SRH guidance. However, their adoption depends on adolescents' perceptions of competence, privacy, usefulness, and social presence. This study will explore these perceptions using a scenario-based survey approach and guided by the Technology Acceptance Model (TAM) (Davis, 1989) and the Unified Theory of Acceptance and Use of Technology (UTAUT) (Venkatesh et al., 2003). The research will examine cognitive and social influences on chatbot adoption. Participants will respond to hypothetical chatbot interaction scenarios simulating SRH consultations. Quantitative measures will assess perceptions across five dimensions—trust, competence, privacy concerns, usefulness, and social presence—while qualitative responses will provide deeper insights into attitudes toward AI-driven SRH interventions. This study aims to identify factors influencing chatbot acceptance among Nigerian adolescents and provide actionable recommendations for designing effective digital health tools. Findings will contribute to the growing body of literature on digital health communication and technology adoption in low-resource settings while addressing critical gaps in SRH information access for Nigerian youth.

Research Questions:

1. How do Nigerian adolescents perceive the competence of an AI chatbot in providing accurate and reliable sexual and reproductive health (SRH) information
2. To what extent do Nigerian adolescents perceive the confidentiality of their conversation during interaction with an AI chatbot for sexual and reproductive health (SRH) information?
3. Do Nigerian adolescents perceive an AI chatbot useful for obtaining sexual and reproductive health (SRH) information?
4. To what extent do Nigerian adolescents perceive the interaction with an AI chatbot for sexual and reproductive health (SRH) information as human-like?
5. How do demographic factors (e.g., age, gender, tech adoption level, education) influence Nigerian adolecents of AI chatbots for sexual and reproductive health (SRH) information?

6. How do adolescents describe their overall experience with the simulated SRH AI chatbot?

Keywords: AI Chatbots · Sexual and Reproductive Health · Technology Acceptance Model · Privacy Concerns · Adolescents in Nigeria

References

1. Ajibade, B.L., Oyedele, E.A., Ajayi, A.D., Amoo, P.O.: Recommendations for removing access barriers to effective sexual/reproductive health services (SRHS) for young people in South East Nigeria: a systematic review. Int. J. Sex. Reprod. Health Care **5**(1), 104–114 (2022)
2. Davis, F.D.: Perceived usefulness, perceived ease of use, and user acceptance of information technology. MIS Q. **13**(3), 319–340 (1989)
3. Nmadu, A.G., Mohamed, S., Usman, N.O.: Barriers to adolescents' access and utilization of reproductive health services in a community in north-western Nigeria: a qualitative exploratory study in primary care. Afr. J. Prim. Health Care Fam. Med. **12**(1), e1–e5 (2020)
4. Venkatesh, V., Morris, M.G., Davis, G.B., Davis, F.D.: User acceptance of information technology: toward a unified view. MIS Q. **27**(3), 425–478 (2003)

Contents

AI for Social Good and Public Impact in Marginalized Communities: A Cross-Cultural Framework

Azmine Toushik Wasi[(✉)]

Shahjalal University of Science and Technology, Sylhet, Bangladesh
`azmine32@student.sust.edu`

Abstract. Artificial Intelligence (AI) has the potential to address systemic inequities and contribute to social good, yet marginalized communities—such as racial minorities and low-income populations—remain largely excluded from its benefits. Despite AI's promise in sectors like healthcare, climate resilience, and education, its deployment often reinforces biases and overlooks cultural and infrastructural challenges. This article addresses these issues by proposing a 10-step framework for developing culturally responsive, community-driven AI solutions that prioritize equity and inclusivity. By incorporating ethical, technical, and policy considerations, the framework aims to bridge the gap between AI advancements and the needs of underserved populations. The proposed approach has significant implications for fostering sustainable AI adoption that empowers marginalized communities and drives progress towards the United Nations' Sustainable Development Goals (SDGs).

Keywords: Equitable AI · AI for Social Good (AI4SG) · Algorithmic Bias · Inclusive AI Governance · Community-Driven AI · Sustainable Development Goals (SDGs)

1 Introduction

Artificial Intelligence (AI) presents an unprecedented opportunity to address some of the world's most pressing challenges, offering transformative potential in sectors such as healthcare, education, climate action, and social justice. However, the benefits of AI have not been distributed equitably, leaving marginalized communities—including racial minorities, low-income populations, and people in resource-constrained regions—systematically excluded from its advantages [1, 2]. AI applications in fields like- healthcare, where predictive models are being used for disease diagnosis and treatment recommendations, can unintentionally perpetuate existing racial and socioeconomic biases. For example, healthcare algorithms trained on historically biased datasets may offer less accurate predictions for minority populations, exacerbating disparities in health outcomes [3]. Similarly, in education, AI-driven learning systems have been criticized for failing to adapt to the linguistic, cultural, and infrastructural challenges faced by underserved students. Many of these AI systems assume a one-size-fits-all approach,

Y. Folajimi et al. (Eds.): SIAI 2025, CCIS 2599, pp. 1–12, 2025.
https://doi.org/10.1007/978-3-031-98949-0_1

often neglecting the realities of students from different socioeconomic backgrounds, leading to further exclusion [3, 4]. Moreover, AI tools are often developed without sufficient input from those most affected, resulting in solutions that do not adequately address the needs of marginalized communities. To ensure that AI fulfills its promise for social good (AI4SG), it is crucial to adopt a paradigm shift that emphasizes equity, community-driven innovation, and cultural sensitivity. This means not only developing more inclusive algorithms but also ensuring that the design, deployment, and governance of AI systems prioritize the voices of those who have historically been excluded from technology development.

Fig. 1. Ethical AI4SG in Marginalized Communities: Identifying Problems to Developing Solutions

Despite growing recognition of the role AI can play in achieving the United Nations Sustainable Development Goals (SDGs), significant structural and ethical barriers persist in its equitable implementation. A 2023 UN report reveals that only 15% of SDG targets are on track, with millions of individuals—particularly in low- and middle-income countries (LMICs)—continuing to face pressing challenges like climate vulnerability, food insecurity, and access to healthcare [4]. These populations are often left out of the AI design and decision-making processes, resulting in tools that fail to capture their diverse needs and experiences. As AI technologies advance, there is an increasing need for these systems to be developed with a comprehensive understanding of cultural, social, and economic contexts. Ethical concerns such as algorithmic bias, lack of transparency, and issues of data privacy are exacerbated when these technologies are deployed in environments with weak regulatory oversight or where governance frameworks do not reflect local realities [3]. For instance, when AI systems are deployed in regions with limited access to digital infrastructure, the benefits can be unevenly distributed, reinforcing existing inequalities. Furthermore, AI governance structures that do not involve marginalized communities in decision-making perpetuate power imbalances, as technologies are designed and implemented by those who may not fully understand the specific challenges faced by underserved populations. Achieving true impact in AI4SG requires a deliberate and holistic shift towards human-centered AI, one that integrates ethical considerations, cultural understanding, and community agency into the design, deployment, and monitoring of AI solutions. Only then can AI be leveraged as a tool for real, lasting change, contributing to global equity and justice.

To bridge the existing gaps, this article presents a comprehensive 10-step framework designed to guide the development of culturally responsive and community-driven AI solutions. We identify key challenges for AI4SG and outline a comprehensive framework for this, as illustrated in Fig. 1. Our framework emphasizes key principles such as participatory design, ethical AI governance, and localized capacity-building to foster sustainable and inclusive innovation. By integrating insights from interdisciplinary research, the proposed approach provides actionable pathways to address both technical and socio-political challenges in scaling AI4SG initiatives. Ultimately, the goal is to empower marginalized communities as active stakeholders in AI development, ensuring that these technologies serve as enablers of social progress rather than reinforcing existing inequities.

2 Background

AI for Social Good (AI4SG) initiatives have made notable strides in addressing global challenges across climate action, healthcare, and education, often aligning with the United Nations' Sustainable Development Goals (SDGs) [1, 4]. Projects such as Amnesty International's Troll Patrol have leveraged AI to quantify online abuse against women, showcasing the potential for interdisciplinary efforts to advance gender equality (SDG 5) [1, 5]. Similarly, AI-driven climate informatics has been instrumental in predicting deforestation and optimizing renewable energy deployment, supporting SDG 13 (climate action) [1, 4]. Furthermore, initiatives like JoyNet, developed by Dr. Desmond Patton, highlight how AI can foster digital resilience and amplify positive narratives among marginalized youth [6]. These efforts demonstrate AI's versatility in addressing social challenges; however, their impact remains unevenly distributed, with significant gaps in equity and inclusivity.

A critical examination of existing AI4SG projects reveals several persistent shortcomings. Geographic disparities continue to hinder progress, with only 10% of AI4SG grants allocated to low- and middle-income countries—regions where SDG needs are most pressing [4]. Moreover, AI systems often replicate historical biases, as evidenced by racial disparities in healthcare algorithms that disproportionately disadvantage minority communities [6]. Ethical gaps also persist, with few projects adhering to internationally recognized frameworks such as the OECD Principles on AI, which emphasize fairness, transparency, and human rights [1, 7]. Cultural biases further com- pound these issues, as highlighted by the Global AI Cultures workshop, which found that AI systems frequently reflect Western-centric norms that fail to align with the cultural values and lived experiences of diverse populations [7, 8]. Large language models (LLMs), such as GPT-4, risk erasing local knowledge and perpetuating cultural homogenization, while facial recognition systems have been shown to misidentify people of color at disproportionately high rates due to imbalanced training data [7].

Our work builds upon these insights by addressing the gaps in existing AI4SG efforts through a more holistic, culturally responsive, and community-driven approach. Unlike previous projects that often take a top-down or one-size-fits-all approach, our proposed 10-step framework prioritizes equity, inclusivity, and participatory design, ensuring that marginalized communities are actively engaged in the AI development process. By integrating ethical, technical, and policy considerations from the outset, our work offers a more comprehensive strategy for deploying AI in cross-cultural contexts. This framework not only aims to mitigate biases and ethical concerns but also fosters sustainable innovation that empowers local communities, ultimately bridging the divide between technological advancements and social impact.

3   Identified Challenges

Despite the transformative potential of AI4SG, several persistent challenges hinder its equitable deployment and effectiveness in marginalized communities. These challenges stem from biases in data and design, ethical and governance gaps, cultural insensitivity, and resource inequities, as outlined in Fig. 2. This section explores these challenges in detail.

3.1   Bias in Data and Design

Historical biases embedded in datasets continue to perpetuate systemic racism, sexism, and classism in AI outputs [3, 6]. Many AI systems are trained on data that reflects existing societal inequalities, leading to discriminatory outcomes that disproportionately affect marginalized populations. For instance, predictive policing algorithms trained on biased crime data often reinforce racial profiling, while healthcare AI models trained on predominantly Western datasets fail to generalize to diverse populations. Moreover, the lack of representation within AI development teams exacerbates these biases. With limited inclusion of diverse voices—especially from low-resource regions—AI systems are often designed with assumptions that do not align with the lived realities of marginalized users [9]. This results in solutions that overlook critical socio-cultural factors, rendering them ineffective or even harmful in their intended contexts.

3.2   Ethical and Governance Gaps

A significant portion of AI4SG initiatives—over 70%—remain confined to research and pilot projects rather than scalable, real-world deployments [4]. This gap between research and implementation is further compounded by the absence of robust accountability mechanisms, leading to ethical lapses and potential misuse of AI technologies in vulnerable communities. The fragmented landscape of AI regulations presents another challenge. Conflicting regulatory frameworks, such as the European Union's AI Act (https://artificialintelligenceact.eu/), which emphasizes transparency and fairness, contrast sharply with more surveillance-focused policies in countries like China [7]. This regulatory divergence complicates efforts to establish universal ethical standards, making it difficult to implement AI solutions that are both globally applicable and locally relevant.

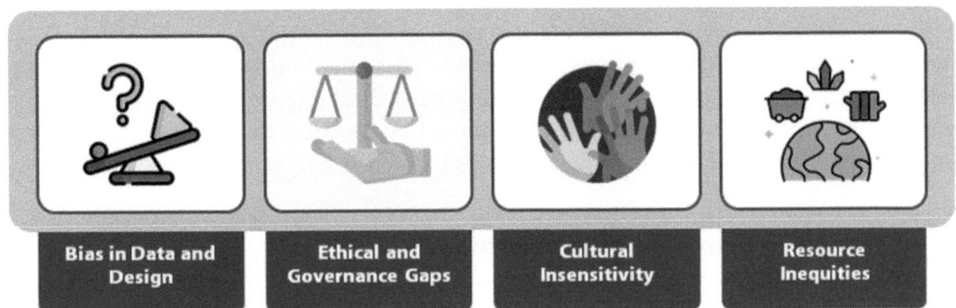

Fig. 2. Identified Challenges in Developing Ethical AI4SG in Marginalized Communities

3.3 Cultural Insensitivity

AI systems are often designed with high-resource environments in mind, neglecting linguistic diversity, infrastructural limitations, and local customs [8]. This one-size fits all approach leads to significant cultural misalignment, as seen in the scarcity of AI literacy materials available in indigenous languages such as Yoruba, which limits the participation of Nigerian communities in digital education initiatives [8]. Furthermore, the failure to incorporate local cultural knowledge into AI systems results in tools that may conflict with traditional values and social norms. For example, chatbots designed for Western healthcare settings may not align with culturally embedded health-seeking behaviors in rural African or South Asian communities, thereby reducing their effectiveness and acceptance.

3.4 Resource Inequities

Marginalized regions often face severe data poverty, lacking the necessary infrastructure to collect and process high-quality datasets [1, 10]. This data scarcity limits the ability to develop AI models that accurately reflect the unique challenges faced by these communities, leading to gaps in service provision and decision-making. Additionally, the allocation of private capital for AI4SG initiatives remains skewed. Investments are heavily concentrated in sectors such as autonomous vehicles and energy, which primarily benefit high-income regions. In contrast, SDG-related challenges such as Zero-Hunger (SDG 2) and Quality Education (SDG 4) receive limited funding, leaving critical social issues unaddressed [4]. Bridging this gap requires targeted investment strategies that prioritize underserved populations and essential social services.

Addressing these challenges demands a multi-faceted approach that includes equitable data practices, inclusive AI design, harmonized governance frameworks, culturally aware AI development, and targeted resource allocation. Our proposed framework aims to tackle these issues by embedding equity, inclusivity, and participatory engagement at every stage of AI development.

4 Framework for Ethical AI4SG in Marginalized Communities

Developing ethical AI4SG solutions in marginalized communities requires a holistic approach that accounts for cultural, social, and infrastructural realities. Our proposed framework consists of ten interconnected steps aimed at ensuring inclusivity, transparency, and sustainability in AI implementation, as illustrated in Fig. 3.

4.1 Community-Driven Problem Definition

A critical first step in designing AI solutions for marginalized communities is ensuring that these communities are actively involved in defining the problems AI aims to address. Traditional top-down approaches often fail to capture the nuanced challenges and lived realities of marginalized populations, resulting in ineffective or even harmful interventions. To counter this, AI developers should employ participatory design methods such as community-led focus groups, co-design workshops, and ethnographic studies to better understand local needs [6]. These methods enable stakeholders to voice their concerns, aspirations, and priorities, fostering a sense of ownership and trust in the AI systems being developed. Culturally accessible methods like comic boards, storytelling, and visual narratives can bridge AI literacy gaps and ensure inclusivity, particularly in communities with low digital literacy [9]. Engaging local leaders and trusted intermediaries can further facilitate meaningful discussions, ensuring that AI solutions align with local cultural values and socio-economic conditions.

4.2 Ethical and Cultural Audits

Ethical and cultural audits are essential to identify potential risks, biases, and cultural misalignments before AI deployment. AI technologies often reflect the biases embedded in their training data, which can perpetuate systemic inequalities and exacerbate existing social divisions. Ethical audits should encompass a comprehensive assessment of data sources, algorithmic decision-making processes, and potential societal impacts using established frameworks such as the Montreal Declaration for Responsible AI [11]. Cultural audits, on the other hand, should focus on evaluating AI's alignment with local customs, values, and beliefs through frameworks like Hofstede's cultural dimensions [12]. For example, an AI system designed for Western healthcare practices may not adequately capture the holistic health approaches prevalent in non-Western societies. Regular consultations with cultural anthropologists, sociologists, and local stakeholders can help mitigate potential cultural insensitivities [17]. Additionally, pre-deployment pilots and feedback loops should be integrated to refine AI models based on community input. For regional knowledge, AI Models can be evaluated on benchmarks, such as, WorldBench [13] and INCLUDE [14].

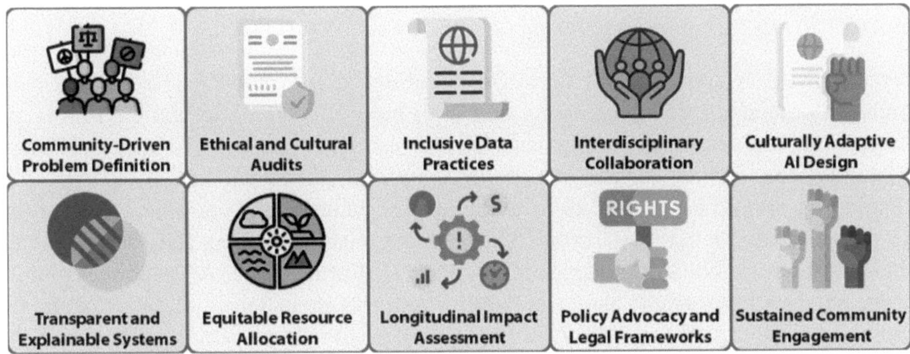

Fig. 3. Components of Our Framework for Ethical AI4SG in Marginalized Communities

4.3 Inclusive Data Practices

Ensuring that AI systems are built on diverse, representative datasets is fundamental to achieving fairness and reducing bias. Many AI systems deployed in marginalized regions suffer from "data poverty," where there is an absence of high-quality, contextually relevant data. Addressing this challenge requires collaboration with local NGOs, community organizations, and academic institutions to collect diverse datasets that reflect the social, economic, and cultural realities of the region. Participatory data collection methods, such as citizen science initiatives and mobile-based data gathering, empower communities to contribute their data and insights. Federated learning approaches can also be employed to enable decentralized data processing while preserving privacy and security, addressing concerns around data ownership and misuse [10]. In addition to quantitative data, qualitative insights from interviews and storytelling should be incorporated to provide a more holistic understanding of local contexts. Efforts should also be made to document data collection methodologies transparently to ensure accountability and replicability [1].

4.4 Interdisciplinary Collaboration

AI4SG initiatives must integrate knowledge and expertise from multiple disciplines to create robust and contextually appropriate solutions. Collaboration between AI researchers, social scientists, policymakers, and domain experts ensures that AI systems account for technical, social, and ethical considerations. Such interdisciplinary efforts foster a deeper understanding of societal needs, ethical concerns, and practical implementation challenges. The "AI Commons" model (https://ai-commons.org/), which encourages collaboration between stakeholders from different sectors, offers a blueprint for fostering interdisciplinary engagement. For example, partnerships between public health officials and AI developers can lead to the creation of more effective disease surveillance tools that are both technically sound and socially acceptable. Engaging local practitioners who have firsthand knowledge of community needs can further enhance the design and deployment of AI solutions. Additionally, cross-sector collaborations help bridge gaps in technical expertise and infrastructure availability, ensuring the sustainability and scalability of AI interventions [15].

4.5 Culturally Adaptive AI Design

Developing AI systems that are sensitive to cultural contexts and infrastructural constraints is crucial for their acceptance and long-term effectiveness. Many AI tools are designed for high-resource settings and fail to consider linguistic diversity, technological limitations, and local customs prevalent in marginalized regions. To address this, AI solutions should be tailored to local languages, dialects, and communication styles using techniques such as low-resource natural language processing (NLP) models and culturally relevant content generation [14, 16, 17]. Furthermore, AI interfaces should be designed with simplicity and accessibility in mind, considering factors such as literacy levels and digital familiarity. Leveraging alternative data sources such as satellite imagery and mobile sensors can help compensate for the lack of traditional IoT infrastructure in resource-constrained settings. Community co-design practices should also be incorporated to ensure that AI systems align with local values and traditions, ultimately enhancing user trust and adoption.

4.6 Transparent and Explainable Systems

Transparency and explainability are vital for building trust and ensuring that AI systems are accountable to the communities they serve. Many marginalized communities have a limited understanding of how AI systems operate, which can lead to distrust and reluctance to adopt AI-driven solutions [10]. Providing clear, multilingual explanations of algorithmic decisions in culturally appropriate formats, such as visual aids and audio explanations, can enhance transparency. Explainability techniques, such as saliency maps and counterfactual explanations, can be incorporated to help users understand AI-driven recommendations in critical domains such as healthcare and social welfare. Policymakers and AI developers should prioritize the creation of regulatory frameworks that mandate transparency in AI deployment, ensuring that communities have access to information regarding data usage, decision-making processes, and avenues for redress in cases of harm or bias.

4.7 Equitable Funding and Resource Allocation

One of the key barriers to the success of AI4SG initiatives in marginalized communities is the unequal distribution of financial and technical resources. Most AI funding is concentrated in commercial sectors such as autonomous vehicles and finance, leaving critical areas such as public health and education underfunded. To address this, AI developers and advocates should push for funding models that prioritize social impact and align with the United Nations' Sustainable Development Goals (SDGs) [1]. Governments, philanthropic organizations, and corporate social responsibility programs must collaborate to provide grants and resources specifically targeted at low- and middle-income countries. Furthermore, capacity-building initiatives should accompany funding efforts to ensure that local communities have the skills and knowledge to leverage AI solutions effectively.

4.8 Longitudinal Impact Assessment

Measuring the long-term impact of AI interventions is essential to ensure their sustainability and prevent unintended consequences. Traditional short-term evaluation metrics, such as accuracy and efficiency, often fail to capture the broader social and economic effects of AI solutions. Instead, AI4SG projects should adopt comprehensive assessment frameworks that track their contributions across multiple SDGs, such as poverty reduction, gender equality, and environmental sustainability. Regular impact assessments, conducted through community feedback and third-party audits, can help identify potential areas of improvement and unintended negative consequences. The insights gained from these assessments should be used to iteratively refine AI systems and ensure their alignment with evolving community needs and priorities.

4.9 Policy Advocacy and Legal Frameworks

Robust policy frameworks and regulatory mechanisms are crucial for ensuring accountability and protecting the rights of marginalized communities in AI deployment. AI4SG initiatives should advocate for hybrid global-local governance models that balance universal ethical principles with region-specific regulations. This includes adapting existing legal frameworks, such as the EU's GDPR, to accommodate the unique socio-cultural and economic conditions of non-Western regions. Governments and civil society organizations should work together to establish clear guidelines for data protection, consent, and algorithmic accountability, ensuring that AI systems operate transparently and ethically.

4.10 Sustained Community Engagement

Continuous engagement with communities post-deployment is essential for the iterative improvement and long-term success of AI systems. Establishing community review boards and participatory governance structures allows for ongoing feedback and adaptation of AI solutions. Digital platforms and offline forums should be utilized to facilitate communication between AI developers and community members, ensuring that their voices remain central to the development process. This sustained engagement fosters trust, promotes social inclusion, and enhances the resilience of AI4SG initiatives in marginalized settings.

5 Discussion

5.1 Overcoming Implementation Barriers

The successful deployment of ethical AI4SG in marginalized communities faces numerous implementation challenges, primarily arising from systemic power imbalances and scalability concerns. Tech companies and government institutions often retain disproportionate control over AI-driven initiatives, sidelining the voices of local communities in critical decision-making processes. To address this, there is a pressing need to cede decision-making authority to marginalized populations, ensuring they have ownership over AI solutions that directly impact their lives. The Community AI Solutions Model

(CASM) offers a promising example, where youth-led design initiatives have demonstrated increased acceptance and effectiveness of AI interventions by integrating local knowledge and contextual insights. Furthermore, scalability remains a major hurdle, as ethical AI development can be resource-intensive and technically demanding. Leveraging automated machine learning (AutoML) tools can significantly reduce development costs while maintaining ethical rigor by streamlining the model-building process and enabling non-experts to participate in AI development. Additionally, skills-based volunteering programs can help bridge the resource gap by enabling AI experts to contribute their knowledge to grassroots initiatives, fostering a more inclusive AI ecosystem.

5.2 Ethical Dilemmas

Implementing AI systems in marginalized communities requires a delicate balance between technical utility and ethical considerations. One of the most pressing dilemmas is the trade-off between bias and utility. While complex AI models such as deep neural networks often achieve higher accuracy, they may introduce opacity and perpetuate biases present in training data. In contrast, simpler and more interpretable models like logistic regression can enhance fairness and transparency at the expense of predictive performance. This trade-off highlights the need for context-aware decision-making, where ethical considerations are prioritized based on the specific social and cultural implications of AI interventions. Another key challenge lies in reconciling global and local ethical frameworks. For instance, an AI ethics framework tailored to the Nigerian context may prioritize linguistic inclusion and accessibility, addressing local needs such as multilingual support and dialect adaptation. In contrast, European Union-style privacy laws focus heavily on stringent data protection measures, which may not align with the immediate priorities of resource-constrained communities. This divergence underscores the importance of developing region-specific ethical guidelines that reflect local values while adhering to basic human rights principles.

5.3 Future Directions

Looking ahead, the evolution of AI4SG in marginalized communities is poised to benefit from advancements in generative AI and decolonial AI perspectives. Generative AI, if designed with cultural sensitivity, has the potential to democratize content creation and amplify marginalized voices. Initiatives such as K-Datasets, which curate culturally relevant data, can empower underrepresented communities by enabling them to produce locally meaningful content across various domains, including education, healthcare, and digital storytelling. Furthermore, the concept of decolonial AI is gaining traction, emphasizing the need to center indigenous knowledge systems in AI design and decision-making processes. The Global AI Cultures workshop has highlighted the potential of incorporating traditional wisdom and locally rooted epistemologies into AI systems, ensuring that technological advancements align with the lived realities of marginalized populations. These forward-looking approaches provide a roadmap for developing AI solutions that are not only technically robust but also socially and culturally inclusive.

Moving forward, interdisciplinary collaborations and participatory research methodologies will play a critical role in advancing these efforts and fostering sustainable, community-driven AI ecosystems.

6   Conclusion

To effectively harness AI for social good, we must shift from techno-solutionism to a justice-oriented approach that emphasizes community agency, cultural humility, and interdisciplinary collaboration. While existing AI4SG efforts have made strides, they often overlook critical issues like bias, cultural insensitivity, and resource inequities. Our proposed framework addresses these gaps by empowering marginalized communities as co-designers and ensuring that AI solutions are culturally relevant and ethically sound.

The challenges of AI deployment in low-resource regions require a concerted effort to tackle power imbalances, invest in local capacity-building, and develop policies prioritizing ethical AI over profit. Striking the balance between fairness and utility, while considering both global and local ethical standards, is essential for avoiding the reinforcement of existing inequities.

Looking ahead, embracing generative and decolonial AI paradigms can democratize access to technology and amplify marginalized voices. By aligning AI4SG initiatives with the SDGs and prioritizing inclusivity, we can transform AI from a tool of exclusion into a force for global equity. This framework calls for collaboration across stakeholders to ensure AI is used responsibly to advance social justice and human dignity for all.

Acknowledgement. We are grateful to HerWILL Inc. and Computational Intelligence and Operations Laboratory (CIOL) for all kinds of support and guidance in the work.

References

1. Cowls, J., Tsamados, A., Taddeo, M., Floridi, L.: A definition, benchmark and database of AI for social good initiatives. Nat. Mach. Intell. **3**(2), 111–115 (2021). https://doi.org/10.1038/s42256-021-00296-0
2. Toma˘sev, N., et al.: Ai for social good: unlocking the opportunity for positive impact. Nat. Commun. **11**(1) (2020). https://doi.org/10.1038/s41467-020-15871-z
3. Ajanaku, D.: How artificial intelligence impacts marginalized communities. Published on 26 Jan 2022, Accessed 23 Jan 2025 (2022). https://sites.law.berkeley.edu/thenetwork/2022/01/26/how-artificial-intelligence-impacts-marginalized-communities/12
4. Medha Bankhwal, A.B.R.R.A.V.H. Michael Chui: AI for social good: Improving lives and protecting the planet. McKinsey Company. Published 10 May 2024, Accessed 23 Jan 2025 (2024). https://sites.law.berkeley.edu/thenetwork/2022/01/26/how-artificial-intelligence-impacts-marginalized-communities/
5. Lahoti, S.: Troll patrol report: amnesty international and element AI use machine learning to understand online abuse against women. Packt. Accessed 23 Jan 2025 (2018)
6. Patton, D., Mohammed, L.: Mitigating AI harms on marginalized populations: key takeaways from our AI and marginalized communities discussion. Accessed 23 Jan 2025 (2024). https://alltechishuman.org/all-tech-is-human-blog/mitigating-ai-harms-marginalized-communities

7. Sethu Sankaranarayanan, L.: The global AI framework: navigating challenges and societal impacts. AI amp; SOCIETY (2024). https://doi.org/10.1007/s00146-024-02070-3

8. Qadri, R., et al.: Global AI cultures. In: ICLR 2024 Workshops (2024). https://openreview.net/forum?id=gw8aq110vD

9. Li, T., Iacobelli, F.: Purposeful AI. In: Companion Publication of the 2023 Conference on Computer Supported Cooperative Work and Social Computing. CSCW 2023 Companion, pp. 563–565. ACM, New York, NY, USA (2023). https://doi.org/10.1145/3584931.3606954

10. Latham, A., Crockett, K.: Towards trustworthy AI: raising awareness in marginalized communities. In: 2024 International Joint Conference on Neural Networks (IJCNN), pp. 1–8 (2024). https://doi.org/10.1109/IJCNN60899.2024.10650957

11. Morand´ın-Ahuerma, F.: Montreal declaration for responsible AI: 10 principles and 59 Recommendations

12. Hofstede, G.: Cultural dimensions in management and planning. Asia Pac. J. Manag.ement $1(2)$, 81–99 (1984). https://doi.org/10.1007/bf01733682

13. Moayeri, M., Tabassi, E., Feizi, S.: Worldbench: quantifying geographic disparities in llm factual recall. In: Proceedings of the 2024 ACM Conference on Fairness, Accountability, and Transparency. FAccT 2024, pp. 1211–1228. ACM, New York, NY, USA (2024). https://doi.org/10.1145/3630106.3658967

14. Romanou, A., et al.: INCLUDE: evaluating multilingual language understanding with regional knowledge. In: The Thirteenth International Conference on Learning Representations (2025). https://openreview.net/forum?id=k3gCieTXeY

15. Patton, D.U., Frey, W.R., McGregor, K.A., Lee, F.-T., McKeown, K., Moss, E.: Contextual analysis of social media: the promise and challenge of eliciting context in social media posts with natural language processing. In: Proceedings of the AAAI/ACM Conference on AI, Ethics, and Society. AIES 2020, pp. 337– 342. ACM, New York, NY, USA (2020). https://doi.org/10.1145/3375627.3375841

16. Putri, R.A., Haznitrama, F.G., Adhista, D., Oh, A.: Can LLM generate culturally relevant commonsense QA data? Case study in Indonesian and Sundanese. In: Al-Onaizan, Y., Bansal, M., Chen, Y.-N. (eds.) Proceedings of the 2024 Conference on Empirical Methods in Natural Language Processing, pp. 20571–20590. ACL, Miami, Florida, USA (2024). https://doi.org/10.18653/v1/2024.emnlp-main.1145

17. Anik, M.A., Rahman, A., Wasi, A.T., Ahsan, M.M.: Preserving cultural identity with context-aware translation through multi-agent AI systems (2025). https://arxiv.org/abs/2503.04827

Balancing Power and Ethics: A Framework for Addressing Human Rights Concerns in Military AI

Mst Rafia Islam[1] and Azmine Toushik Wasi[2(✉)]

[1] Independent University, Dhaka, Bangladesh
[2] Shahjalal University of Science and Technology, Sylhet, Bangladesh
azmine32@student.sust.edu

Abstract. AI has made significant strides recently, leading to various applications in both civilian and military sectors. The military sees AI as a solution for developing more effective and faster technologies. While AI offers benefits like improved operational efficiency and precision targeting, it also raises serious ethical and legal concerns, particularly regarding human rights violations. Autonomous weapons that make decisions without human input can threaten the right to life and violate international humanitarian law. To address these issues, we propose a three-stage framework (Design, In Deployment, and During/After Use) for evaluating human rights concerns in the design, deployment, and use of military AI. Each phase includes multiple components that address various concerns specific to that phase, ranging from bias and regulatory issues to violations of International Humanitarian Law. By this framework, we aim to balance the advantages of AI in military operations with the need to protect human rights.

Keywords: Military AI · Human Rights · AI Ethics · Power and Politics

1 First Section

AI has advanced significantly in recent years, resulting in a wide range of applications in both the civilian and military sectors. The military, which is continually seeking innovations for more effective, faster, and stronger technology or weapons, sees AI as a perfect answer to address these needs [1, 2]. AI technologies offer many advantages, such as increased operational efficiency, precision targeting, and reduced human casualties [3, 4]. However, these advancements come with profound ethical and legal concerns, particularly regarding potential human rights violations [4–6].

Use of AI in military applications raises serious concerns [2, 7, 8]. Autonomous weapons systems, capable of making life-or-death decisions without human intervention, threaten the right to life and violate international humanitarian law. While AI-enhanced surveillance systems improve data collection, they can lead to widespread privacy violations and unnecessary monitoring of individuals [2]. Additionally, inherent biases in AI systems may result in discriminatory practices, exacerbating existing disparities and violating principles of equality, non-discrimination, and human rights [4, 5].

Y. Folajimi et al. (Eds.): SIAI 2025, CCIS 2599, pp. 13–17, 2025.
https://doi.org/10.1007/978-3-031-98949-0_2

Motivated by these concerns, in this work, we propose a novel three stage framework to address human rights issues in the design, deployment, and use of military AI, with each aspect consisting of multiple components. Our work aims to propose a comprehensive framework for evaluating human rights violations by military AI, addressing both technical and statutory aspects. By examining the ethical implications and regulatory challenges, this framework seeks to provide a balanced approach to harnessing the benefits of AI in military operations while safeguarding human rights.

2 Framework

As shown in Fig. 1, this framework addresses human rights concerns in the Military AI lifecycle across three key phases: Design, Deployment, and Use, with each phase having several components. Below, we discuss each phase and its components in detail.

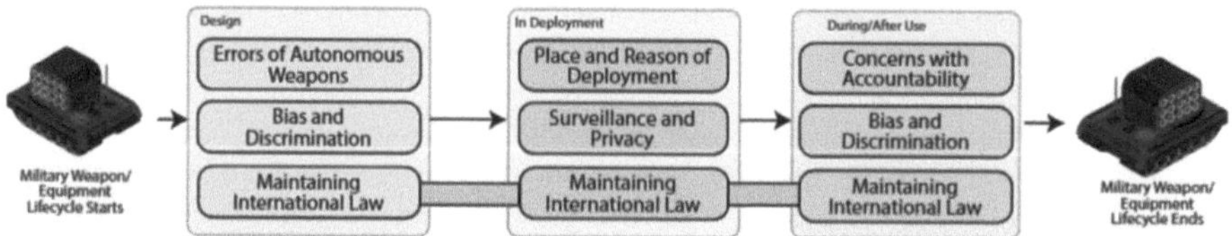

Fig. 1. Our framework for addressing human rights concerns in military AI.

2.1 Design Phase

Design phase involves creating military AI systems, focusing on algorithms and functionalities. It sets the foundation for AI behavior, ensuring it operates correctly, without bias and discrimination.

Targeting Errors of Autonomous Weapons. AI-powered autonomous weapons systems can make life-and-death decisions without human intervention [9]. AI systems can make errors in identifying and targeting individuals, leading to unintended casualties and injuries among civilians [10]. Implementing ethical guidelines in the design and development of military AI systems can ensure compliance with human rights standards. Moreover, incorporating human oversight in AI decision-making processes ensures accountability and ethical compliance. Human-in-the-loop (HITL) [11] systems allow human operators to intervene and override AI decisions when necessary.

Bias and Discrimination. AI systems can perpetuate or even exacerbate existing biases, leading to discriminatory practices in targeting or profiling based on race, ethnicity, or other characteristics [12]. By developing algorithms that detect and mitigate biases in AI systems can prevent discriminatory outcomes. This not only enhances the fairness of AI applications but also helps in building public trust and ensuring compliance with ethical standards.

2.2   In Deployment Phase

In Deployment phase involves integrating military AI systems into specific environments, focusing on their usage while emphasizing ethical considerations and legal compliance.

Surveillance and Privacy. AI-driven surveillance systems can infringe on individuals' privacy rights, leading to unauthorized monitoring and data collection [13]. Implementing strong data privacy and security measures to protect sensitive information collected and processed by military AI systems. This includes encryption, access controls, and regular security assessments.

Deployment of the Weapon. To ensure maximum efficacy while minimizing collateral damage [14], the automated weapon system must be deliberately positioned in a location that maximizes operational range and precision. In order to prevent unintended harm, this entails a detailed investigation of the terrain, potential targets, and civilian areas [15]. Continuous monitoring and changes are required to ensure its efficacy and safety. Additionally, real-time data analysis and feedback loops should be integrated to adapt to dynamic battlefield conditions, ensuring that the system remains responsive and minimizes risks to non-combatants [10].

2.3   During/After Use Phase

During/After Use phase encompasses the actual operation of military AI systems during conflicts or missions. It assesses the AI's performance in real-time, monitoring for accountability and compliance with human rights and legal standards.

Concerns with Accountability. Transparency of AI decision-making processes complicates accountability, making it difficult to attribute responsibility for human rights violations [16]. When AI systems make decisions, it creates a gap in justice and responsibility, as it becomes challenging to hold any individual or entity accountable for wrongful actions [17]. By ensuring AI systems are transparent and their decision-making processes are explainable, accountability issues can be resolved. Ethical guidelines, such as the Asilomar AI Principles [18], emphasize the importance of transparency, accountability, and human oversight in the development and deployment of military AI. Adhering to these principles can help mitigate the violations of human rights.

Bias and Discrimination. AI systems can perpetuate or worsen existing biases, leading to discriminatory practices in targeting or profiling based on race, ethnicity, or other characteristics [13]. To mitigate these biases post-deployment, continuous monitoring and auditing of AI systems are essential. Implementing feedback loops to identify and correct biased outcomes can ensure fairness [19]. This approach addresses biases and helps build public trust in AI technologies by demonstrating a commitment to ethical standards and accountability.

2.4 Violation of International Humanitarian Law

Deployment of AI in military operations poses serious risks to International Humanitarian Law (IHL), which aims to protect non-combatants. Autonomous weapons can make decisions without human oversight, leading to indiscriminate attacks that violate the principle of distinction [10]. AI targeting decisions may also compromise proportionality, as these systems might not effectively balance military advantage with civilian harm [15]. The lack of transparency and accountability in AI decision- making complicates compliance with IHL, hindering the investigation and attribution of unlawful actions. Adherence to IHL requires strong ethical guidelines, oversight mechanisms, and international cooperation to prevent violations and uphold the laws of war.

3 Challenges and Future Work

Several challenges may hinder the proper implementation of our framework. There is no global agreement on the ethical use of military AI. Groups like the Campaign to Stop Killer Robots push for a ban, but leading AI nations resist binding commitments. Accountability is also complex, as it's hard to determine who is responsible when AI systems cause harm—creators, commanders, or the AI itself. Additionally, AI evolves quickly, outpacing legal systems and creating oversight gaps that could be exploited.

Future research should focus on establishing international agreements and fostering collaboration between engineers, ethicists, and legal experts to keep pace with AI advancements.

Acknowledgements. We are grateful to Computational Intelligence and Operations Laboratory (CIOL) for all kinds of support and guidance in the work.

References

1. Fischer, S.-C.: Military AI applications: a cross-country comparison of emerging capabilities, pp. 39–55. Springer (2022). https://doi.org/10.1007/978-3-031-11043-64
2. Grand-Cl´ement, S.: Artificial intelligence beyond weapons: application and impact of AI in the military domain. United Nations Institute for Disarmament Research (UNIDIR), Geneva (2023)
3. Rashid, A.B., Kausik, A.K., Al Hassan Sunny, A., Bappy, M.H.: Artificial intelligence in the military: an overview of the capabilities, applications, and challenges. Int. J. Intell. Syst. **2023**, 1–31 (2023) https://doi.org/10.1155/2023/8676366
4. Murray, D.: Adapting a human rights-based framework to inform militaries' artificial intelligence decision-making processes. St. Louis University Law J. **68** (2024)
5. Bradley, C., Wingfield, R., Metzger, M.: National artificial intelligence strategies and human rights: a review. Technical report, Global Partners Digital and Global Digital Policy Incubator, Stanford Cyber Policy Center (April 2020). Research assistance from Madeline Libbey and Amy MacKinnon
6. EU: Eu artificial intelligence act. Technical report, European Union (2024). http://data.eur opa.eu/eli/reg/2024/1689/oj

7. Marwala, T.: Militarization of AI Has severe implications for global security and warfare. United Nations University, UNU Centre (2023). https://unu.edu/article/militarization-ai-has-severe-implications-global-security-and-warfare

8. Adam, D.: Lethal AI weapons are here: how can we control them? Nature **629**(8012), 521–523 (2024). https://doi.org/10.1038/d41586-024-01029-0

9. Christie, E.H., Ertan, A., Adomaitis, L., Klaus, M.: Regulating lethal autonomous weapon systems: exploring the challenges of explainability and traceability. AI Ethics **4**(2), 229–245 (2023). https://doi.org/10.1007/s43681-023-00261-0

10. Asaro, P.: On banning autonomous weapon systems: human rights, automation, and the dehumanization of lethal decision-making. Int. Rev. Red Cross **94**(886), 687–709 (2012). https://doi.org/10.1017/s1816383112000768

11. Mosqueira-Rey, E., Hern´andez-Pereira, E., Alonso-R´ıos, D., Bobes-Bascar´an, J., Fern´andez-Leal, A.: Human-in-the-loop machine learning: a state of the art. Artif. Intell. Rev. **56**(4), 3005–3054 (2023)

12. Gordon, F.: Virginia Eubanks (2018) automating inequality: how high-tech tools profile, police, and punish the poor. New York: Picador, St martin's press. Law, Technology and Humans, pp. 162–164 (2019) https://doi.org/10.5204/lthj.v1i0.1386

13. Feldstein, S.: The global expansion of AI surveillance vol. 17. Carnegie Endowment for International Peace Washington, DC (2019)

14. Stein, T.: Collateral damage, proportionality and individual international criminal responsibility. In: International Humanitarian Law Facing New Challenges, pp. 157–161. Springer, Berlin, Heidelberg (2007)

15. Crootof, R.: The killer robots are here: Legal and policy implications. Cardozo Law Rev. **36**, 1837 (2014)

16. Novelli, C., Taddeo, M., Floridi, L.: Accountability in artificial intelligence: what it is and how it works. AI Soc. **39**(4), 1871–1882 (2024)

17. Pasquale, F.: The Black Box Society: The Secret Algorithms That Control Money and Information. Harvard University Press (2015). https://doi.org/10.4159/harvard.9780674736061

18. Future of Life Institute: Asilomar AI Principles. Coordinated by FLI and developed at the Beneficial AI 2017 conference. (2017). https://futureoflife.org/open-letter/ai-principles/

19. Raji, I.D., et al.: Closing the AI accountability gap: defining an end-to-end framework for internal algorithmic auditing (2020) 2001.00973 [cs.CY]

Amharic Language Audio Data Search Engine

Zemenfes Hailemariam Gebremedhin[1]([✉]) [iD] and Mulubrhan Abebe Nerea[2] [iD]

[1] Addis Ababa Institute of Technology (AAiT), Addis Ababa University, Addis Ababa, Ethiopia
zemenfeshailemariam@gmail.com
[2] Gothubrug, Sweden
mulubrhan-abebe.nerea@student.hv.se

Abstract. The generation of audio files from various sources, including the internet and social media, has increased significantly in the rapidly expanding digital landscape. It is difficult to efficiently access specific spoken words from this vast collection of Amharic audio data. To address this, we propose a novel method that combines Text-Based Spoken Term Detection (STD) with models. Our methodology includes speech segmentation using Pydub, the development of an ASR model, and the implementation of keyword-based STD. The ASR model successfully transcribes audio files, allowing meaningful keywords to be extracted for more accurate and frequent search queries. An analysis of 37 audio files reveals that the sentence error rate (SER) is 91.7 percent (33 of 36 sentences have errors), and the word error rate (WER) is 98.3 percent (285 of 290 words have errors). It improves search accuracy and efficiency for specific spoken terms, significantly enhancing search capabilities for users of Amharic multimedia resources. However, the study emphasizes the need for a larger dataset to improve transcription capabilities and reduce errors, with the potential to revolutionize Amharic audio search engines and empower users in accessing precise information from Amharic audio data, ultimately transforming how we interact with and use Amharic audio resources.

Keywords: ASR (Automatic Speech Recognition) · STD (Spoken Term Detection) · WER (Word Error Rate) · SER (Search Error Rate

1 Introduction

1.1 Background

In the rapidly expanding digital landscape, a vast number of audio files are being generated from diverse sources such as the internet and social media, contributed by individuals and organizations. Among these valuable resources are audio files in Amharic, the official language of Ethiopia (Ado, Gelagay, and Johannessen 2021) (Heinonen 2011) [1, 2]. However, the challenge lies in efficiently searching for specific spoken words within this extensive collection of Amharic audio files.

Traditional audio players lack the essential functionality for word-based audio search, leaving users grappling to extract desired information from the stored Amharic audio data

Y. Folajimi et al. (Eds.): SIAI 2025, CCIS 2599, pp. 18–25, 2025.
https://doi.org/10.1007/978-3-031-98949-0_3

(Lamere and et al. 2003a) [4]. To overcome this obstacle, the development of specialized audio search engines be- comes imperative, capable of matching and retrieving relevant documents exclusively from the database of stored Amharic language audio files. By enabling such advanced audio search capabilities, users can save time and improve their ability to access precise information tailored to their needs.

The primary goal of this study is to address this pressing need by exploring effective methods for searching particular words in audio files recorded in Amharic. The findings hold the potential to revolutionize information retrieval through the creation of cutting-edge search engines specifically de- signed for Amharic multimedia. These advancements will empower individuals, groups, and organizations relying on Amharic audio data, enabling them to seamlessly access pertinent information and identify specific passages within the audio files, ultimately transforming the way we interact with and harness the potential of Amharic audio data.

2 Implementation Detail

2.1 Methodology

The methodology begins with Speech Segmentation, breaking long speech signals into smaller segments using Pydub. These segmented audio chunks are then trained with corresponding text transcriptions via the CMU Sphinx ASR model. Utilizing the Sphinx4 library in Python, we seamlessly align the components of our ASR system. We use a comparison methods of searching techniques but since the target of our search is all spoken words. For specific spoken word searches in Amharic audio, we employ keyword-based Spoken Term Detection (STD), offering efficient and precise speech location within the audio file. Our approach inte- grates tools like CMU Sphinx, Sphinx4, Python, and pydub, enabling effective searching for spoken terms in Amharic audio.

Speech Segmentation is the first step in our methodology, where we divide a long speech signal into smaller segments. This segmentation process, facilitated by the Python package pydub, divides the audio file into frames based on silence, making it more manageable for further analysis. Be- low, we illustrate the different searching techniques (Lamere and et al. 2003b,a; Yadav and et al. 2020; Author 2007) [3, 5, 6].

The creation of an ASR (Automatic Speech Recognition) model follows speech segmentation (Fig. 1).

We train the segmented audio chunks to their corresponding text transcriptions using the CMU Sphinx, an open-source ASR development toolkit [3]. We utilize the Sphinx4 speech recognition library in conjunction with Python to build and interface our ASR system (Lamere and et al. 2003a; Yadav and et al. 2020) [5, 6], resulting in a seamless alignment of various components.

To find specific spoken speech within an audio file using Amharic text as a keyword, we investigate various searching methods, including STD (Spoken Term Detection), QBE STD (Query By Example STD), and keyword detection. Given our research's primary goal of identifying specific spoken words in an audio file using Amharic text, we choose keyword-based Spoken Term Detection (STD) as our preferred searching method. This strategy enables efficient and precise location of the desired speech fragments within the audio file.

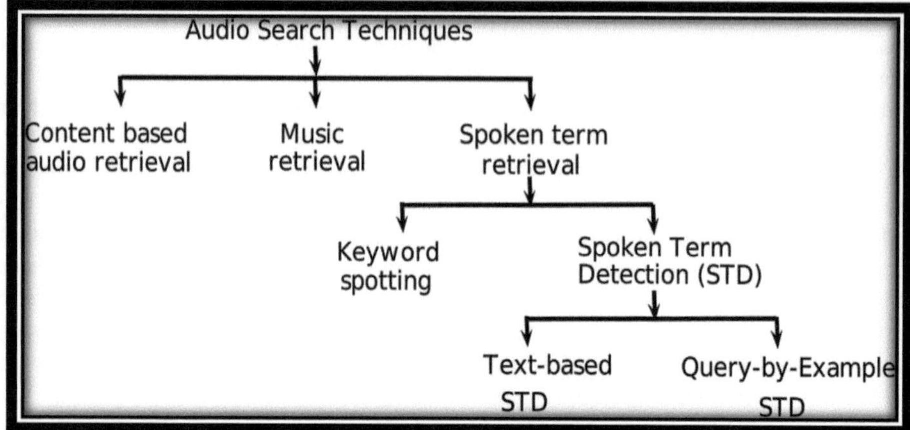

Fig. 1. Techniques of Audio searching

Our research's central methodology revolves around the integration of various tools and techniques, including CMU Sphinx, Sphinx4, Python, and Pydub. This integration process empowers us to efficiently and precisely search for specific spoken terms in the Amharic audio files, contributing to the development of advanced search engines tailored to Amharic multimedia.

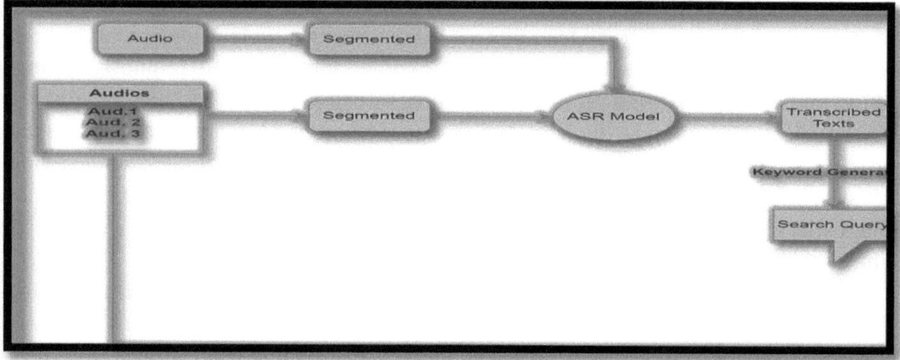

Fig. 2. General Flow Chart

2.2 Experiment

In this process, two audio files, "The Searching Audio" and the "List of Audio to be Searched," are transcribed using an ASR model to generate keywords. The transcriptions of the "List of Audio" are used to create an index, facilitating efficient searches. The generated keyword acts as a search query to identify and retrieve audio files from the "List of Audio" that match the specified term. These retrieved audio files are then presented as the output. This approach streamlines the process of finding and accessing relevant audio content, as depicted in work flow the performance shows all the WER, CER, Precision, recall and F-1 score (Figs. 3 and 4).

```
C:\CMU\searchquery\Scripts\python.exe C:\CMU\searchquery\Lib\Amharic_search_final.py
Audio to search:
ፋሬስና ዛሬን ወለደ ፋሬስም አስሮምም ወለደ አስሮምም አራምን ወለደ አራምም አሚናዳብን ወለደ አሚናብም ፍሬአቸሁም ግን ወለደ
Keywords:
['ወለደ', 'አስሮምም', 'ፋሬስና', 'ዛሬን', 'ፋሬስም']
Match found in:
 C:\CMU\Amharic\list\MWCone15.wav
Transcript: በበሪሉን ምርኮ ጊዜ ኢኮንያንንና ወንድሞቹን ወለደ
Match found in:
 C:\CMU\Amharic\list\MWCone16.wav
Transcript: ከበሪሉንም ምርኮ አለ አየ ከከቡ ያለ ሰላትያልን ወለደ ሰላትያልም ዘሩባቤልን ወለደ


Process finished with exit code 0
```

Fig. 3. Transcriptions of Audio

```
Keywords: ወለደ, አለ, አስሮምም, ነአሶንን, አይተን
Transcript: አይተን ይሁኑ አለ ወለደ አለ ሰምቶ ከዓለም አስሮምም ወለደ አስሮምም ነአሶንን ወለደ አራምም አሚናዳብን ወለደ አሚናብም ነአሶንን ወለ

Match found in: C:\CMU\Amharic1\list\MWCone1.wav
Transcript: የሟሪል ልጅ አለ ልጅ የኢየሱስ ክርስቶስ ትውልድ መጽሐፍ
Match found in: C:\CMU\Amharic1\list\MWCone2.wav
Transcript: አለ ይስሐቅን ወለደ ይስሐቅም ያዕቆብን ወለደ ያዕቆብም
Match found in: C:\CMU\Amharic1\list\MWCone4.wav
Transcript: አይተን ይሁኑ አለ ወለደ አለ ሰምቶ ከዓለም አስሮምም ወለደ አስሮምም ነአሶንን ወለደ አራምም አሚናዳብን ወለደ አሚናብም ነአሶንን ወለ
```

Fig. 4. Searching of Audio from List of Audio

3 Results and Conclusion

Pydub is used to segment a 12-minute input audio wave file into smaller chunks during the training phase. Following that, an ASR model is created using a language model with 2737 sentences and 6231 words, as well as n-gram language models (1-gram: 6231, bigram: 14731, trigram: 16079). The acoustic model was trained using 100 audio samples, which lasted 12 minutes and included 247 phones and 1882 words. A phonetic dictionary with 373 phone numbers is also used. The trained ASR model is evaluated on 37 different wave files. The results show that the sentence error rate (SER) is 91.7% (33 of 36 sentences have errors) and the word error rate (WER) is 98.3% (285 of 290 words have errors). These evaluation metrics are critical in determining the ASR model's performance and accuracy for the given task. As illustrated in Figure 2, The list of audios is correctly matched by the complete transcription of the audios and the search using a most frequent keyword-base as the search query. The small dataset used for training is one of the ASR model's main limitations. To improve transcription capabilities and reduce errors use a large dataset (Fig. 5 and Table 1).

Source Code — https://github.com/zemenfes-afk/Amharic- Audio-search-engine

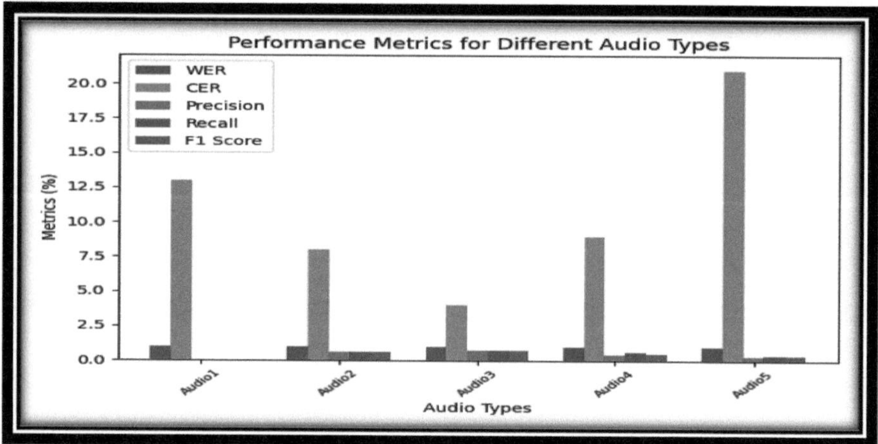

Fig. 5. Performance Metrics of Different Audios

Table 1. Summary of the Chapters

Aspect	Description
Audio Seg-mentation	Segmentation of a 12-min audio file into smaller chunks using PyDub
ASR Model Creation	The ASR model is created with a language model comprising 2737 sentences and 6231 words, along with n-gram language models (1-g, bigram, and trigram). The acous- tic model is trained with 100 audio samples, containing 247 phones, 1882 words, and a phonetic dictionary
Evaluation	The trained ASR model is evaluated on 37 audio files. The evaluation results indicate a Sentence Error Rate (SER) of 91.7% (33 out of 36 sentences have errors) and a Word Error Rate (WER) of 98.3% (285 out of 290 words have errors), along with F-1 score, precision, and recall results
Matching Audio Transcrip-tions	The code successfully matches audio files with their corresponding transcriptions and conducts keyword-based searches
Recommen-dation	The text suggests that the ASR model's performance is limited due to a small dataset and recommends using a larger dataset for training to enhance accuracy

Algorithm 1:   Discounted N-Gram Model with Backoff

Input: Corpus of sentences or text, optional word file for extra vocabulary
Parameters:

- `discount mass`: Discount factor for smoothing
- `case`: Case handling (e.g., lower, upper)
- `add start`: Whether to add start and end tokens to sen- tences
- `norm`: Token normalization flag (e.g., remove punctua- tion)
- `verbose`: Verbosity flag for logging

Output: ARPA-formatted   N-Gram model

1: Initialize the ARPA model object with parameters
  2: **if** word file is provided **then**
3:   Read words from the word file and add to the model
  4: **end if**
  5: **if** corpus file or input text is provided **then**
6:   Read and process the corpus, breaking it into sen- tences
  7: **for** each sentence in the corpus **do**
8:    Tokenize and normalize the sentence
  9: **if** add start is true **then**
10:     Add start and end tokens to the sentence
  11: **end if**
12:    Count 1-grams, 2-grams, and 3-grams in the sen- tence
  13: **end for**
  14: **end if**
15: Calculate unigram, bigram, and trigram probabilities with smoothing
  16: **for** each unigram in the corpus **do**
17:   Calculate the probability using discounting method
  18: **end for**
  19: **for** each bigram in the corpus **do**
20:   Calculate bigram probability using discounting
  21: **end for**
  22: **for** each trigram in the corpus **do**
23:   Calculate trigram probability using discounting
  24: **end for**
25: Output the ARPA-formatted N-Gram model
  26: **return** ARPA model

Algorithm 2: Keyword Extraction and Audio Segmentation

Input: Transcript of audio, list of stopwords
    Parameters: None
Output: List of top 5 keywords
  1: Convert transcript to lowercase
  2: Split the transcript into words
    3: **for** each word in the transcript **do**
    4: **if** word is not in stopwords **then**
  5:       Remove punctuation from the word
  6:       Add the word to a frequency count dictionary
    7: **end if**
    8: **end for**
  9: Sort the words by frequency in descending order
 10: Select the top 5 words as keywords
   11: **return** the list of top 5 keywords

Algorithm 3: Audio Transcription and Segmentation

Input: Audio file, acoustic model, language model, phonetic dictionary
    Parameters: None
Output: Transcription of the audio
  1: Initialize the decoder with given acoustic model, language model, and dictionary
  2: Open the audio file for reading
  3: Start audio processing
    4: **while** audio buffer has data **do**
  5:    Process the audio buffer with the decoder
    6: **end while**
  7: End the audio processing
  8: Retrieve the transcription from the decoder
  9: **if** transcription exists **then** 10: **return** the transcription 11: **else**
    12:    **return** None
    13: **end if**

<u>Algorithm 4: Search and Match Transcripts Algorithm</u>

Input: audio_files, keywords, acoustic_model_path, language_model_path, phonetic_dictionary_path
Parameters: None
Output: List of matching (audio_file, transcript) pairs

1: Initialize empty matches list
2: for each audio_file in audio_files do
3:     transcript = transcribe_audio(audio_file, acoustic_model_path, language_model_path, phonetic_dictionary_path)
4:     if transcript exists then
5:         for each keyword in keywords do
6:             if keyword (case-insensitive) in transcript (case-insensitive) then
7:                 add (audio_file, transcript) to matches
8:                 break keyword loop
9:             end if
10:        end for
11:    end if
12: end for
13: return matches

Acknowledgments. We extend our deepest gratitude to the Ethiopian Artificial Intelligence Institute and Dr. Roza Tsegay Aga, Director of the Division, for her exceptional support as advisor and project leader in the field of Natural Language Processing Her guidance, expertise, and encouragement have been in- valuable to the success of this research.

We are truly grateful and delighted for her leadership and unwavering commitment throughout this journey, Thank you for reading these instructions carefully. We look forward to receiving your electronic files!

References

1. Ado, D., Gelagay, A., Johannessen, J.: The languages of Ethiopia. Grammatical Sociolinguistic Aspects Ethiopian Lang. **48**, 1 (2021)
2. Author, F.: Automatic speech recognition for an under-resourced language—Amharic. In: Eighth Annual Conference of the International Speech Communication Association (2007)
3. Heinonen, K.: Consumer activity in social media: managerial approaches to consumers' social media behavior. J. Consum. Behav. **10**(6), 356–364 (2011)
4. Lamere, P., et al.: The CMU SPHINX-4 speech recognition system. In: IEEE International Conference on Acoustics, Speech, and Signal Processing (ICASSP 2003), vol. 1, Hong Kong (2003a)
5. Lamere, P., et al.: The CMU SPHINX-4 speech recognition system. In: IEEE International Conference on Acoustics, Speech, and Signal Processing (ICASSP 2003), vol. 1, Hong Kong (2003b)
6. Yadav, R., et al.: A model for recapitulating audio messages using machine learning. In: 2020 International Conference for Emerging Technology (INCET), vol. 75. IEEE (2020)

Scaling Inclusive AI Education Through Feedback Analytics in Underserved Contexts: Empowering Learners for the Generative AI Era

Yetunde Folajimi[1]([✉]), Leonidas Deligiannidis[1], Salem Othman[1], Shawren Singh[2], and Chika Yinka-Banjo[3]

[1] School of Computing and Data Science, Wentworth Institute of Technology, 550 Huntington Avenue, Boston, MA 02115, USA
`{folajimiy,deligiannidisl,othmans1}@wit.edu`
[2] Department of Information Science, University of South Africa, Preller Street, Muckleneuk Ridge, Pretoria 0002, South Africa
`singhs@unisa.ac.za`
[3] Department of Computer Science, University of Lagos, University Road, Akoka, Lagos 0002, Nigeria
`cyinkabanjo@unilag.edu.ng`

Abstract. This paper presents the design, deployment, and evaluation of a two-day hybrid workshop aimed at scaling inclusive access to generative AI education in under-served contexts. Hosted in Nigeria and delivered through a partnership between institutions in Africa and North America, the DSAI Workshop integrated hands- on training in prompt engineering and big data analysis with panel discussions on AI ethics and policy. Leveraging a feedback analytics framework, we analyzed both structured and open-ended participant responses to evaluate learning outcomes, engagement patterns, and program impact. Our findings show that 90% of participants rated the experience highly beneficial, with strong gains in technical skill development, regulatory awareness, and networking. A word cloud and thematic coding of qualitative feedback corroborated these outcomes. The participant cohort—primarily beginners and intermediates—benefited from scaffolded instruction, hybrid access, and local relevance. We contribute a replicable, feedback-driven model for inclusive AI capacity building and offer policy recommendations for educational equity in the era of large language models. The results suggest that short-format, data-informed interventions can serve as effective entry points for broader AI literacy efforts across the Global South.

Keywords: Artificial Intelligence Education · Feedback Analytics · Inclusive Learning · Resource-Limited Settings · Generative AI

L. Deligiannidis, S. Othman, S. Singh and C. Yinka-Banjo—These authors contributed equally to this work.

1 Introduction

The rapid evolution of Artificial Intelligence (AI), particularly with the rise of generative models such as GPT-4 [1] and Claude [2], is transforming every sector, including education, healthcare, agriculture, and governance. While these advancements offer unprecedented opportunities, they also highlight a widening gap in access to AI education between high-income and low-resource regions [3]. In many under-served contexts—especially across sub-Saharan Africa—the barriers to AI education include limited infrastructure, lack of skilled educators, and minimal exposure to cutting-edge tools [4]. Bridging this divide is not merely a technological challenge, but a socio-educational imperative that demands inclusive strategies and scalable interventions.

1.1 Motivation

The current trajectory of AI innovation risks excluding a significant portion of the global population from meaningful participation in shaping and benefiting from AI systems. Generative AI, in particular, is reshaping creative industries, coding practices, customer service, and decision support systems. Yet, access to foundational training in areas like prompt engineering, large language models (LLMs), and ethical AI remains concentrated in elite institutions and well-resourced countries [5]. As such, inclusive AI education initiatives are urgently needed to ensure that emerging economies, such as Nigeria, are not left behind in the generative AI revolution.

1.2 Overview of the DSAI Workshop

To address this gap, a hybrid two-day training workshop on Data Science and Artificial Intelligence (DSAI) was organized from November 28–29, 2024, at a university-based robotics laboratory in Nigeria. The event was the result of a cross-institutional collaboration between academic and policy-focused institutions in North America and West Africa. Designed to foster inclusive AI literacy, the workshop combined technical sessions, keynote addresses, and panel discussions to introduce participants to both practical applications and policy-relevant dimensions of generative AI. Key highlights of the workshop included hands-on training in prompt engineering and conversational AI, the application of big data analytics using PySpark and large language models, and a panel discussion on AI regulations and ethics. Importantly, the workshop was delivered through a hybrid model to accommodate both remote and in-person participants, expanding accessibility to a broader audience across Nigeria.

As shown in Fig. 1, the event was organized across two days, blending keynote talks, technical training, and collaborative discussions into a cohesive learning experience.

1.3 Goals

The core aim of the DSAI Workshop was to empower learners—especially students, early-career researchers, and technology professionals from underserved backgrounds—with the foundational knowledge and skills needed to engage with the generative AI ecosystem. By leveraging hands-on instruction, localized delivery, and contextual relevance, the program sought to:

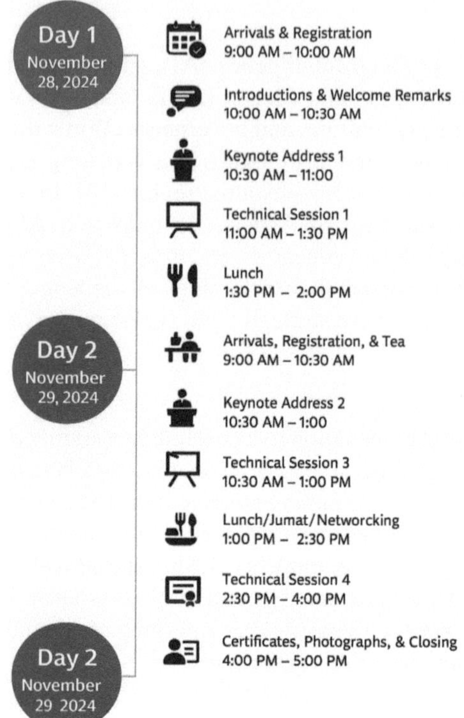

Fig. 1. Two-day visual schedule of the DSAI Workshop showing keynotes, technical sessions, and networking activities across in-person and remote formats.

- Lower the entry barrier into AI fields for participants in resource-limited settings.
- Introduce scalable frameworks for teaching advanced AI topics such as LLMs and reinforcement learning.
- Encourage ethical awareness and critical thinking around AI deployment in Africa.
- Promote international collaboration and access to research and scholarship opportunities.

1.4   Contributions of This Paper

This paper presents a case study of the DSAI Workshop and offers four main contributions:

1. It documents a replicable model for delivering hybrid AI education in underserved regions using a feedback-driven, learner-centered design.
2. It analyzes participant demographics, attendance patterns, and engagement using structured survey data and feedback analytics.
3. It discusses lessons learned from deploying prompt engineering and big data analytics training in a low-resource setting.
4. It offers policy and pedagogical recommendations for scaling inclusive AI education across Africa and other Global South contexts.

Through this work, we aim to contribute to the growing literature on AI for social good [6, 7] and advocate for data-driven strategies that support equitable AI capacity building globally.

2 Background and Related Work

The transformative power of Artificial Intelligence (AI) is increasingly evident across diverse domains, yet access to quality AI education remains uneven globally. Numerous reports and studies point to a critical need for inclusive strategies that extend AI learning opportunities to marginalized and underserved populations, particularly in Africa and the Global South [3, 4, 8]. This section synthesizes prior efforts and situates the DSAI workshop within three intersecting strands of research: inclusive AI education, feedback-driven learning, and hybrid capacity-building models.

2.1 Inclusive AI Education in Underserved Contexts

Recent years have seen a growing acknowledgment of the disparities in AI education and research access. A 2023 report by UNESCO emphasized that developing nations must be supported in AI skills development to avoid deepening the digital divide [9]. Similarly, the AI4D Africa program, funded by the International Development Research Centre (IDRC), has highlighted regional capacity gaps in AI teaching, data literacy, and ethical awareness [10].

Several initiatives have attempted to address this gap. For example, the African Master's in Machine Intelligence (AMMI) has trained hundreds of graduate students in AI across West Africa [11]. Likewise, online platforms like Coursera, fast.ai, and DataCamp have democratized access to AI training but remain limited by language, internet connectivity, and contextual relevance [12]. The DSAI workshop builds upon these efforts by offering localized, hybrid, and hands-on learning, explicitly targeting infrastructural limitations in Nigerian contexts.

2.2 Feedback Analytics for Scalable Learning

Feedback-driven education—enabled by digital tools and analytics—has emerged as a central approach to improving student engagement and learning outcomes [13]. In AI education, adaptive learning platforms such as Code.org, AutoML platforms, and intelligent tutoring systems are integrating feedback loops to personalize learning paths [14, 15].

Recent research has underscored how formative feedback, learner reflections, and real-time assessment tools can improve knowledge retention, especially in short-format learning experiences [16]. For example, Zhang et al. (2023) developed an adaptive reinforcement learning engine for personalized AI coursework, showing higher engagement rates among underrepresented students [17]. The DSAI workshop adopted similar principles by collecting exit feedback and applying that data to assess participant satisfaction, skill acquisition, and future learning needs.

2.3 Hybrid Delivery Models and AI Capacity Building

The COVID-19 pandemic accelerated interest in hybrid education models combining in-person and virtual components [18]. In low-resource settings, hybrid learning must overcome challenges such as inconsistent electricity, limited bandwidth, and lack of devices [19]. Despite this, research from ICT4D and EdTech communities suggests that context-aware hybrid models can significantly improve inclusivity and engagement when designed with local constraints in mind [20].

Capacity-building programs such as Deep Learning Indaba and Data Science Nigeria have successfully delivered hybrid workshops, mentorships, and research grants to foster a new generation of African AI leaders [21, 22]. The DSAI workshop aligns with these initiatives but distinguishes itself through its direct use of feedback analytics and its emphasis on hands-on generative AI tools, including prompt engineering and conversational AI.

2.4 Gaps in the Literature

While the literature supports the potential of inclusive and feedback-driven AI education, few empirical case studies have documented implementations in hybrid, low-resource environments. Most existing studies are either theoretical or focused on online-only interventions. This paper addresses that gap by reporting a real-world, hybrid AI education initiative designed specifically for underserved Nigerian learners, drawing on feedback analytics to inform scalable, data-driven pedagogy.

3 Workshop Design and Implementation

The DSAI Workshop was structured as a two-day hybrid training and panel event aimed at bridging the AI education divide in Nigeria through localized, hands-on learning. This section presents the instructional design, institutional collaboration, curriculum focus, delivery formats, and participant profile of the program.

3.1 Institutional Collaboration and Venue

The workshop was jointly organized by three institutions from the United States and Nigeria:

- A U.S.-based university's School of Computing and Data Science
- A Nigerian federal university's Department of Computer Science
- A national policy and research institute focused on science and technology management

The workshop took place at a robotics laboratory on a major Nigerian university campus, selected for its computing infrastructure and accessibility within Lagos State.

3.2 Curriculum Design and Content

The curriculum was developed to balance technical depth, practical engagement, and policy awareness. It was structured into:

- Two keynote addresses by academic and policy leaders
- Four technical sessions on applied AI topics
- A panel discussion on ethical and regulatory aspects of AI
- An information session focused on research and international collaboration opportunities

Topics emphasized prompt engineering, conversational AI, big data analysis using PySpark, and the ethical considerations of deploying generative AI technologies in Africa.

3.3 Hybrid Delivery Approach

To maximize access, the event was delivered in a hybrid format:

- **In-person**: 47 participants attended physically at the host university.
- **Remote**: 176 participants joined via Zoom, including participants from across Nigeria and the diaspora.

All technical sessions were recorded for asynchronous access, and interactive tools such as live polls and moderated Q&A enhanced remote engagement.

3.4 Participant Profile and Selection

A total of 375 applications were received. Selection was based on expressed interest, diversity of academic backgrounds, and ability to commit to full participation. The final cohort reflected a mix of:

- Undergraduate and graduate students, faculty members, and working professionals
- Participants from over 30 academic institutions and tech hubs across Nigeria
- Skill levels ranging from beginner to intermediate in Python programming

A pre-event survey revealed that while over 70% were familiar with ChatGPT, fewer than 10% had prior experience with models such as DALL·E, BERT, or Mid-Journey. These findings guided the scaffolding of hands-on content and supplementary materials.

3.5 Infrastructure and Resource Strategy

The hybrid model was designed with resource-constrained contexts in mind:

- Materials were distributed ahead of the event to reduce real-time bandwidth demands
- Presentations were recorded and shared via low-bandwidth-friendly formats
- Certificates of participation were issued digitally following verified engagement
- Session slides and prompt templates were compiled into an open-access resource pack

This design showcased how inclusive pedagogy, thoughtful planning, and hybrid delivery can work together to scale AI education without relying on high-cost infrastructure.

4   Participant Engagement and Inclusion Strategy

Achieving broad and meaningful participation was a central design principle of the DSAI Workshop. Given the socio-technical challenges faced by learners in underserved contexts—ranging from infrastructural limitations to unequal exposure to emerging technologies—our strategy emphasized inclusive outreach, accessibility, and tailored engagement. This section outlines how participants were recruited, supported, and engaged during the event.

4.1   Application and Selection Process

The workshop attracted a total of 375 applicants from across Nigeria and the diaspora. Applications were evaluated based on motivation, institutional diversity, and interest in generative AI topics. A total of 223 participants (59.5%) were accepted, comprising 47 in-person attendees and 176 remote participants.

The selection criteria prioritized inclusivity across:

- Academic status (e.g., undergraduates, graduate students, faculty, professionals)
- Geographic diversity, especially applicants outside major tech hubs
- Expressed interest in both technical and policy aspects of AI

To ensure equitable participation, a balance was maintained between first-time learners and those with moderate prior exposure to AI tools.

4.2   Analytical Approach

Structured feedback (Likert and multiple-choice responses) was analyzed using descriptive statistics and visualized using Python's matplotlib and pandas libraries. Open-ended responses were processed using a two-stage qualitative analysis. First, frequent word patterns were extracted using basic natural language processing (tokenization, stopword filtering, and frequency counts) to generate word clouds. Second, responses were inductively coded into thematic categories using a grounded theory approach [23], allowing key patterns such as technical empowerment, ethical awareness, and desire for continuity to emerge organically. No machine learning models were applied, as the dataset size and scope did not warrant automated clustering or sentiment analysis.

4.3   Pre-Workshop Engagement and Preparedness

To bridge the varied experience levels among participants, applicants were asked to self-report their familiarity with key generative AI models during registration. This allowed the organizers to assess baseline competencies and adapt the curriculum accordingly (Fig. 2).

Among the accepted participants, familiarity with ChatGPT stood out:

- 59.2% were *very familiar*
- 33.1% were *somewhat familiar*
- Only 7.7% reported *no familiarity*

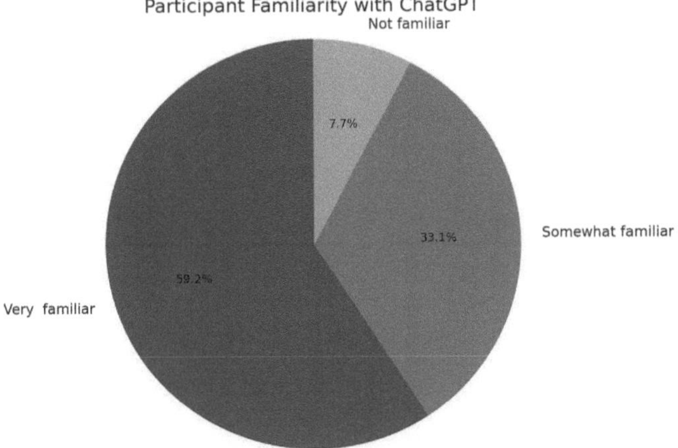

Fig. 2. Self-reported participant familiarity with ChatGPT. A majority had prior exposure, reflecting the tool's widespread public awareness.

While ChatGPT was the most widely recognized generative model, participants' exposure to other tools was significantly more limited. As shown in Fig. 3, familiarity with foundational transformer models like BERT and T5, or visual generative tools like MidJourney and StyleGAN, was much lower. Over 70% of respondents indicated no familiarity with models such as DALL·E, PaLM, and StyleGAN.

This disparity in model familiarity validated the decision to scaffold the workshop's technical sessions from foundational concepts. Introductory walkthroughs, context-setting slides, and live demonstrations were prioritized, especially for models not well-known to the majority of attendees.

4.4 Inclusive Design for Hybrid Participation

A hybrid model was adopted to extend access beyond Lagos and reduce the digital divide. Key inclusion strategies included:

- **Remote Participants:** Provided with Zoom access links, recordings, and low-bandwidth alternatives.
- **In-Person Participants:** Offered dedicated lab space, technical support, and printed resource materials.
- **Language and Accessibility:** All session instructions used clear, non-technical English and avoided jargon.
- **Time-Zone Coordination:** The schedule accommodated both West African and Eastern Standard Time zones to include facilitators and attendees from different regions.

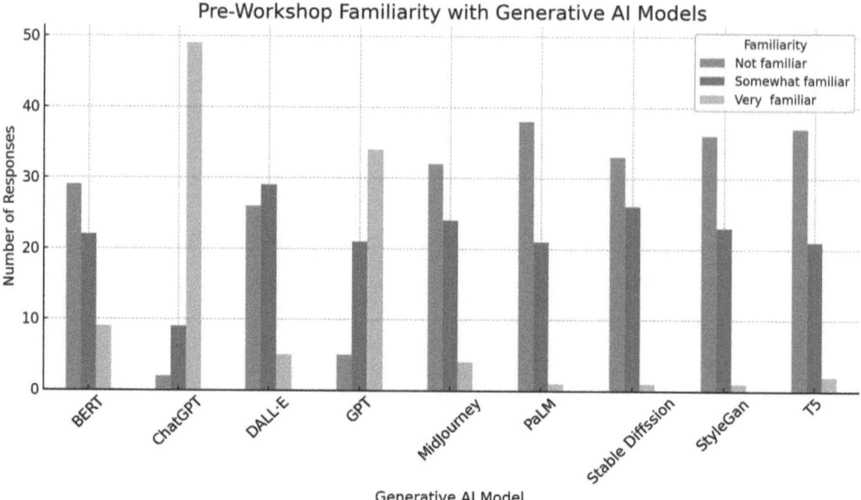

Fig. 3. Pre-workshop familiarity with various generative AI models. Familiarity dropped significantly outside of mainstream tools such as ChatGPT, revealing educational gaps in foundational and visual models.

4.5 Engagement Tools and Communication Channels

To facilitate interaction and feedback:

- Daily reminders and calendar invites were used to reinforce attendance
- Live chat and moderated Q&A supported real-time interaction during sessions
- A dedicated email helpline was set up for technical and participation support

4.6 Attendance Verification and Certification

Participants were eligible for a digital certificate of completion upon verified attendance. Attendance was validated through:

- Zoom login tracking for remote participants
- On-site registration logs for in-person attendees
- Post-session surveys and a final feedback form

This structured approach helped maintain high levels of session completion and post-event engagement, especially among remote participants who might otherwise have dropped off due to technical or scheduling issues.

5 Feedback Analytics and Evaluation

To evaluate the workshop's effectiveness and guide future implementations, structured and open-ended feedback was collected from participants. This section summarizes key insights across benefit ratings, skill development, session value, and qualitative reflections.

5.1 Participant Satisfaction and Workshop Value

Participants rated the workshop's overall benefit on a scale of 1 (least beneficial) to 5 (most beneficial). As shown in Fig. 4, over 90% rated the workshop a 4 or 5, reflecting high satisfaction and alignment between the content and participant expectations.

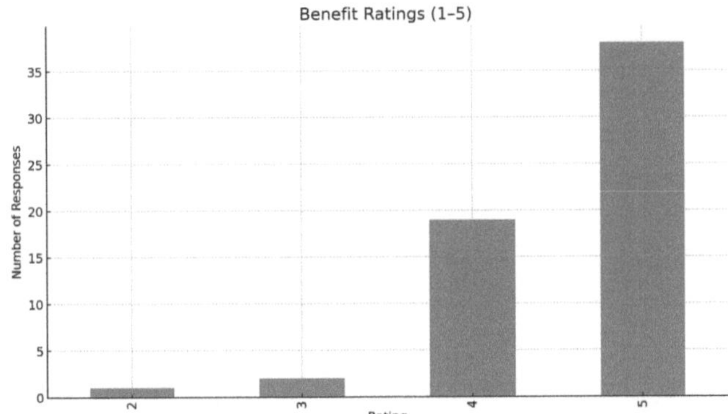

Fig. 4. Distribution of workshop benefit ratings on a scale from 1 (least beneficial) to 5 (most beneficial). A majority rated the event as highly beneficial.

5.2 Reported Benefits and Learning Outcomes

Participants selected multiple outcomes they believed they gained from the workshop. Figure 5 presents the most frequently reported benefits. A large majority cited gains in technical skills and awareness of AI regulations and policies. Networking and exposure to research and fellowship opportunities also featured prominently.

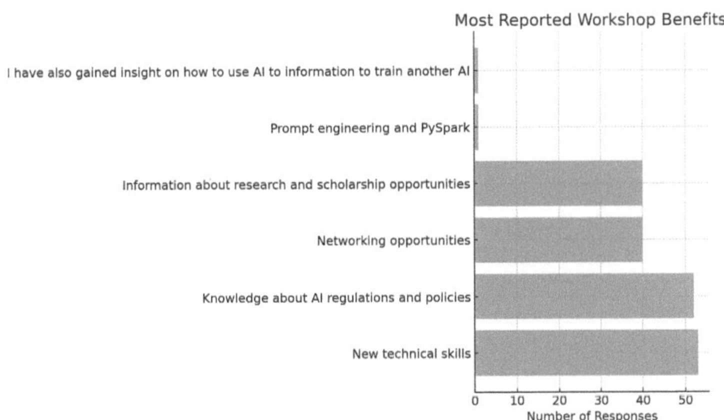

Fig. 5. Most commonly reported benefits of attending the DSAI Workshop. Technical skills and AI policy knowledge were the most cited outcomes.

5.3 Python Skill Levels and Engagement Readiness

Understanding participants' prior skill levels provided insight into the accessibility and impact of the curriculum. As shown in Fig. 6, nearly half of participants identified as beginners, while 45% were intermediate users. Only a small fraction self-identified as advanced.

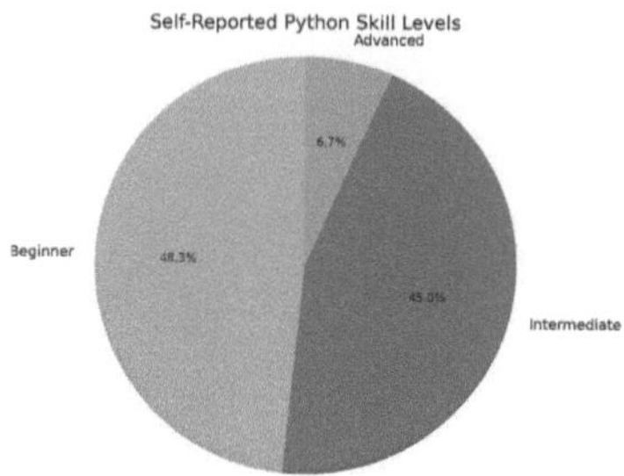

Fig. 6. Participant self-reported Python programming skill levels. Most learners identified as beginner or intermediate.

This distribution confirms the workshop's relevance to early-career learners and justifies the scaffolded instructional approach adopted for technical sessions.

5.4 Qualitative Insights and Thematic Patterns

To complement quantitative responses, several open-ended questions were analyzed to extract recurring themes in participant reflections. Responses addressed what participants found most impactful, suggestions for improvement, final comments, and general observations.

A combined word cloud (Fig. 7) was generated from these multiple fields. Dominant terms such as *"workshop," "training," "prompt," "engineering," "learning," "python,"* and *"session"* highlight the strong emphasis on technical content and hands-on instruction. Expressions of appreciation, such as *"thank"* and *"impactful"*, further indicate positive participant sentiment. Recurring references to *"time"* and *"make"* also suggest calls for extended sessions or deeper content.

This visualization supports thematic patterns previously identified: participants valued practical exposure, desired more time for exploration, and expressed enthusiasm for continuing their learning journeys. Such insights are essential for designing future iterations that are both high-impact and responsive to learner needs.

5.5 Evaluation Summary

The triangulated data—quantitative ratings, structured benefit selection, and qualitative text—paint a clear picture: the DSAI Workshop effectively engaged a technically diverse audience, delivered relevant and accessible content, and fostered interest in further AI learning. The feedback analytics also underscored the importance of iterative design and structured reflection for future capacity-building initiatives in AI education.

Fig. 7. Combined word cloud from multiple open-ended feedback questions. Recurring terms reflect participant engagement with hands-on technical sessions, AI applications, and suggestions for future improvement.

6 Insights and Lessons Learned

The implementation and evaluation of the DSAI Workshop provided not only a successful training experience for participants but also critical operational and pedagogical insights. This section distills lessons learned across five key domains: curriculum design, accessibility, engagement, infrastructure, and scalability.

6.1 Designing for Technical Diversity

One of the most important takeaways was the need to design with a broad range of technical proficiencies in mind. As shown in Fig. 6, nearly half of participants identified as beginners in Python programming, while the remainder were split between intermediate and advanced levels. This required scaffolding the curriculum such that foundational concepts (e.g., syntax, data structures) were reinforced while more complex tasks (e.g., PySpark pipelines, prompt chaining) were introduced gradually. The effectiveness of such scaffolding aligns with literature on differentiated instruction in STEM learning, which emphasizes adapting content to learners' readiness levels and preferred learning modes [24]. Future workshops can build on this by offering tiered tracks or optional preparatory modules for new learners.

6.2 Hybrid Delivery Can Be Inclusive—If Intentional

While hybrid events are becoming more common post-pandemic, delivering a high-impact hybrid experience in a resource-constrained context demands careful planning. Factors such as low-bandwidth accommodations, asynchronous follow-up content, and cross-platform access were crucial in maintaining high engagement among remote attendees.

These findings are echoed in Walton et al.'s work on digital equity, which highlights the importance of infrastructure-aware design for online education in low-income settings [19]. Participants expressed appreciation for the flexibility of the format, but also noted ongoing challenges such as internet reliability—reminders of the infrastructural inequities that still shape digital learning access.

6.3 Feedback Loops Enhance Educational Quality

The integration of feedback analytics throughout the workshop process proved essential. Pre-workshop surveys informed curriculum pacing; real-time polls helped adjust session flow; and post-event surveys, analyzed in Sect. 5, offered evidence of impact and areas for iteration.

This learner-centered feedback loop reinforced a model of continuous improvement and responsiveness. As argued by Zhang et al. (2023), such reinforcement-based feedback systems can significantly improve learner engagement and performance in AI education, particularly among underrepresented groups [17].

6.4 Cross-Institutional Collaboration Drives Capacity Building

The partnership between U.S.-based and Nigerian academic institutions, as well as a national technology policy body, was foundational to the program's success. It provided a blend of technical depth, local relevance, and policy framing that resonated with participants. For example, the inclusion of a panel on AI regulation and ethics contextualized the technical content within societal challenges, a feature several participants cited as impactful.

Such collaborations are key to sustainable AI education models across Africa. Future efforts may benefit from more structured institutional frameworks, such as co-developed syllabi, co-certification programs, and intercontinental mentorship pipelines.

6.5 Limitations and Considerations for Scale

Despite the workshop's successes, several challenges emerged:

- **Session Length:** Many participants noted that the technical sessions were dense and could have benefited from being spread over more days.
- **Practice Time:** A lack of structured time for individual hands-on exercises during the live sessions limited deeper engagement for some.
- **Sustained Learning:** Participants expressed a desire for continuity—through follow-up programs, assignments, or community forums.

While the overall experience was positive, several real-time implementation challenges emerged. Notably, intermittent internet outages affected the pacing of remote sessions, particularly during live coding segments. Some participants also reported difficulty accessing pre-installed software or stable power sources, which are common infrastructural barriers in the region. These disruptions, though mitigated by adaptive facilitation and asynchronous materials, underscore the fragility of digital learning environments in underserved contexts.

Furthermore, maintaining engagement over a two-day intensive format proved challenging, especially for remote participants. Dropout rates increased slightly by the second day, suggesting a need for more interactive pacing or motivational scaffolds. Long-term sustainability remains a key concern—scaling such workshops requires both consistent institutional support and funding models that incentivize continuity beyond one-off interventions.

7 Discussion

The outcomes of the DSAI Workshop underscore both the potential and the challenges of deploying scalable, inclusive AI education in underserved contexts. Drawing on feedback analytics, participation trends, and implementation insights, this section discusses broader implications for the future of generative AI education, policy development, and sustainable capacity building in the Global South.

7.1 Positioning AI Education as a Social Imperative

As large language models (LLMs) continue to reshape knowledge work, access to AI literacy has become a social justice issue as much as a technological one. In regions like sub-Saharan Africa, where historical inequities intersect with emerging digital divides, ensuring that students, educators, and professionals can engage with—and contribute to—the generative AI ecosystem is vital. The DSAI Workshop demonstrates that even with modest infrastructure, it is possible to facilitate meaningful learning through context-aware, hybrid education models.

This aligns with broader international efforts, such as UNESCO's 2023 AI Competency Framework for Educators, which highlights the urgency of embedding AI fluency across disciplines and geographic regions [9]. Our findings reinforce the notion that scalable, equitable AI education must blend technical training with ethical literacy, application relevance, and opportunities for sustained practice.

7.2 Bridging Policy and Practice through Contextualization

One of the distinguishing features of the DSAI Workshop was its integration of both technical and policy-oriented sessions. Feedback data suggest that participants valued the balance between learning tools like ChatGPT and PySpark and engaging with ethical and regulatory questions. This interdisciplinary design mirrors recommendations from global AI ethics frameworks, which stress that AI training must not be dislocated from social context, governance mechanisms, and community engagement [6, 7].

Bringing policymakers, practitioners, and students into the same learning space also promotes bi-directional learning: technologists gain awareness of the societal stakes of their work, while regulators better understand the evolving capabilities of AI systems. In Nigeria and similar contexts, such approaches can strengthen the co-design of future educational and regulatory frameworks.

7.3 The Role of Feedback in Scalable Learning Design

The use of structured feedback mechanisms across the workshop lifecycle proved to be both impactful and practical. From pre-event surveys that shaped content scaffolding to real-time polls that informed instructional pacing, and post-event forms that captured perceptions of value and gaps—feedback analytics served as a lightweight but powerful tool for pedagogical adaptation.

This approach aligns with current research in adaptive education systems, which argues that learning experiences informed by user feedback and behavioral data can improve motivation, retention, and equity outcomes—particularly when deployed in short-format or high-throughput environments [17, 25].

7.4 Toward Sustainable Models for AI Capacity Building

While short-term training workshops offer critical exposure, long-term sustainability requires more systemic integration. Participants indicated strong interest in post- event mentorship, assignments, and communities of practice. This points toward a next phase in inclusive AI education: developing locally anchored, globally connected pipelines that extend beyond workshops into microcredentialing, research internships, and curriculum integration at the university level.

Several promising pathways exist, including open-source AI curricula, bilateral university partnerships, and continent-wide initiatives such as the Deep Learning Indaba or the AI4D Africa network [10, 21]. The DSAI Workshop contributes a replicable model that can complement and inform these broader efforts.

7.5 Limitations and Transferability

While the results from this workshop are encouraging, caution is warranted in generalizing outcomes without contextual adjustments. The success of the hybrid model was supported by existing institutional relationships and access to a university lab facility. In more remote or unconnected areas, alternative delivery mechanisms—such as mobile learning, radio modules, or community learning hubs—may be more appropriate.

Furthermore, feedback-based design, while powerful, still requires regular iteration and infrastructure (e.g., surveys, response tracking, data literacy among facilitators) to be effective. Future implementations should consider how to resource and institutionalize these practices sustainably.

8 Conclusion and Future Work

This paper presented the design, implementation, and evaluation of a two-day hybrid AI education workshop held in Nigeria, with the goal of scaling inclusive access to generative AI tools and knowledge through feedback analytics. Informed by over 375 applications and 223 confirmed participants across in-person and remote formats, the DSAI Workshop demonstrated a replicable, low-cost model for delivering high-impact, hands-on AI training in a resource-constrained context.

8.1 Key Contributions

Our findings contribute to both the academic and practitioner literature on AI education and capacity building in the Global South. Key contributions include:

- A contextualized curriculum focused on prompt engineering, big data analysis with PySpark, and AI policy awareness, delivered through hybrid modalities.
- A feedback-driven instructional model that used participant input to inform design choices before, during, and after the event.
- Empirical insights into participant demographics, technical skill levels, engagement metrics, and perceived benefits of the training.
- A practical case study of international collaboration across academic institutions and national research organizations.

The consistently high satisfaction scores (over 90% rating the event 4 or 5), diverse institutional representation, and rich qualitative feedback suggest that hybrid models, when thoughtfully designed, can overcome infrastructural barriers and meet learners where they are.

8.2 Implications for AI Education Equity

As the generative AI revolution continues to accelerate, education systems worldwide face a dual challenge: ensuring technical fluency among their populations, and doing so in ways that are equitable, localized, and ethically grounded. The DSAI Workshop offers one potential pathway—combining open tools, contextual pedagogy, and collaborative delivery—to bridge these gaps.

The results underscore the importance of feedback analytics, not just as a tool for evaluation but as a pillar for designing adaptive and inclusive education. They also affirm that even short-format interventions, if well-scaffolded and data-informed, can have significant impact in expanding AI literacy in underserved communities.

8.3 Future Work

While the workshop achieved its immediate goals, it also revealed directions for long-term impact. Planned next steps include:

- **Longitudinal Follow-Up:** Conducting a 3–6 month post-workshop survey to track continued learning, application of skills, and research engagement.
- **Micro-Credentialing Pathways:** Developing a modular version of the curriculum with competency-based assessments that can be integrated into local university programs or offered as verified certificates.
- **Open Educational Resources (OER):** Publishing prompt templates, teaching slides, and session recordings as Creative Commons materials to facilitate reuse across Africa and beyond.
- **Community of Practice:** Launching a mentorship group or virtual cohort for continued peer learning, collaboration, and knowledge exchange.
- **Policy Engagement:** Building on the policy panel to initiate workshops and dialogues with regulatory bodies around AI ethics, equity, and educational frameworks.

Through these efforts, we aim to transition from a one-time event to a sustainable model for empowering learners in the generative AI era—an effort that will require not only technology and pedagogy but also intentional collaboration, equity-centered design, and systemic support.

Taken together, these findings inform not only future workshops but broader national and continental strategies for equitable AI literacy. The following policy recommendations synthesize the workshop's insights for decision-makers and educational planners.

> **Policy Recommendations for Inclusive AI Education**
>
> - **Support Hybrid Models:** Governments and educational institutions should invest in low-bandwidth-compatible hybrid learning infrastructure to extend AI education beyond urban centers.
> - **Fund Open Educational Resources (OER):** Policymakers should prioritize funding for the development and localization of open AI curricula, especially those tailored to African socio-economic contexts.
> - **Incentivize Cross-Border Academic Collaboration:** Facilitate institutional partnerships that pair local universities with international AI research labs and departments, including funding for exchange programs and co-certification.
> - **Promote Ethical AI Literacy:** AI education initiatives should be accompanied by national or regional policy guidelines that promote the inclusion of ethics, bias mitigation, and local cultural values.
> - **Leverage Feedback Analytics:** Encourage the adoption of feedback-informed learning design in public sector training programs and university curricula to enhance adaptive learning experiences.

To build on the momentum of this initiative, we call on international funders, academic institutions, and technology partners to invest in scalable, hybrid models for AI education in underserved regions. The success of this workshop demonstrates that inclusive and impactful AI capacity building is not only possible—it is urgently necessary. Collaborative programs that integrate hands-on technical training, ethical literacy, and feedback-driven design offer a pathway to global AI equity in the generative era.

Acknowledgments. The authors gratefully acknowledge the institutional support and collaborative commitment of the partner organizations that made this initiative possible. We thank the National Centre for Technology Management (NACETEM) for its strategic support in advancing AI ecosystem in Nigeria, and the AI and Robotics Laboratory (AIR-LAB) at the Department of Computer Science, University of Lagos, for hosting the workshop and providing technical infrastructure. Special appreciation goes to the AI and Multimedia Systems (AIMS) Research Lab at the School of Computing and Data Science, Wentworth Institute of Technology, for its leadership in curriculum design, session facilitation, and feedback analytics. We also recognize the invaluable contributions of the workshop facilitators, panelists, and volunteers, whose expertise and engagement shaped the success of the event. Finally, we thank the participants for their enthusiasm, critical input, and commitment to continued learning. This work reflects the outcome of a cross-institutional collaboration focused on capacity building in emerging technologies. The views expressed herein are those of the authors and do not necessarily represent those of the supporting institutions.

Ethics Statement.   This study adhered to ethical standards for research involving human participants in educational settings. Participation in the DSAI Workshop and all associated surveys was entirely voluntary. All feedback and data analyzed in this paper were collected anonymously, and no personally identifiable information was stored, shared, or reported.

Participants were informed of the purpose of the data collection and consented to its use for academic reporting and program improvement. The workshop activities were educational in nature and posed no foreseeable risk to participants. No compensation was provided, and participation had no bearing on academic or professional evaluations.

Institutional support was provided by partner universities and research centers in both Nigeria and the United States. No conflicts of interest were declared, and all institutional affiliations are disclosed transparently in the Acknowledgments section.

This work complies with guidelines outlined by the ACM Code of Ethics and the IEEE Code of Conduct regarding responsible computing, educational fairness, and transparency in AI education research.

References

1. OpenAI: Gpt-4 technical report (2023). https://openai.com/research/gpt-4
2. Anthropic: Claude Language Model. https://www.anthropic.com/index/claude (2023)
3. UNESCO: Artificial intelligence and education: Guidance for policy-makers (2021). https://unesdoc.unesco.org/ark:/48223/pf0000376709
4. Ndemo, B., Weiss, T.: Artificial intelligence for development in Africa: Challenges and opportunities. AI Soc. **36**, 753–765 (2021)
5. Vinuesa, R., Azizpour, H., Leite, I., Balaam, M., Dignum, V., Domisch, S., et al.: The role of artificial intelligence in achieving the sustainable development goals. Nat. Commun. **11**(1), 233 (2020)
6. Jobin, A., Ienca, M., Vayena, E.: The global landscape of AI ethics guidelines. Nat. Mach. Intell. **1**(9), 389–399 (2019)
7. Floridi, L., Cowls, J.: Ai4people: an ethical framework for a good ai society. Mind. Mach. **30**(1), 1–24 (2020)
8. Andela: Andela Launches Technical Leadership Program to Upskill African Developers. https://andela.com. https://andela.com/blog/andela-launches-technical-leadership-program (2022)
9. UNESCO: AI Competency Framework for Teachers. https://unesdoc.unesco.org/ark:/48223/pf0000384622 (2023)
10. (IDRC), I.D.R.C.: AI for Development Africa Program. https://www.idrc.ca/en/initiative/artificial-intelligence-development-africa (2021)
11. Mathematical Sciences, A.I.: African Masters in Machine Intelligence (AMMI). https://aims.ac.za/research-centres/ammi/ (2021)
12. Whittaker, M.: Disability, bias, and ai. AI Now Institute Annual Report (2021)
13. Hattie, J.: Visible Learning for Teachers: Maximizing Impact on Learning. Routledge (2012)
14. Roschelle, J., Lester, J.C.: Next-generation digital learning environments: a vision for personalizing learning. Educ. Technol. **59**(1), 46–53 (2019)
15. Zhang, W., Lin, H.: Adaptive learning strategies for scalable ai education. IEEE Trans. Learn. Technol. **15**(3), 281–293 (2022)

16. Lokar, M., Tissenbaum, M., Linn, M.C.: Designing formative feedback for artificial intelligence courses. In: Proceedings of the 53rd ACM Technical Symposium on Computer Science Education (SIGCSE), pp. 121–127. ACM (2022). https://doi.org/10.1145/3478431.3499313
17. Zhang, L., Patel, R.: Reinforced feedback mechanisms for personalized ai education in underrepresented groups. J. Educ. Data Min. **15**(1), 87–106 (2023)
18. Hodges, C., Moore, S., Lockee, B., Trust, T., Bond, A.: The difference between emergency remote teaching and online learning. Educause Review (2020). https://er.educause.edu/articles/2020/3/the-difference-between-emergency-remote-teaching-and-online-learning
19. Walton, M., Hassreiter, S.: Tech access and digital equity in African classrooms. Inf. Technol. Int. Dev. **17**, 56–72 (2021)
20. Blake, E.: Local learning contexts for global technologies. https://ict4d.org.za/2022/12/10/local-learning/ (2022)
21. Indaba, D.L.: Annual report 2022). https://deeplearningindaba.com/2022-annual-report/ (2022)
22. Nigeria, D.S.: Impact Report 2023. https://www.datasciencenigeria.org/impact/ (2023)
23. Charmaz, K.: Constructing Grounded Theory. Sage Publications (2014)
24. Rose, D.H., Meyer, A.: Universal design for learning in stem education: a framework for inclusive pedagogy. J. Postsecondary Educ. Disab. **35**(1), 23–38 (2022)
25. Hwang, G.-J., Xu, W.-H., Wang, Y.-Y.: Trends in adaptive learning: A review of research from 2010 to 2020. J. Educ. Technol. Soc. **23**(4), 68–82 (2020)

Towards Culturally Inclusive Knowledge Dissemination Using AI Agents

Mahfuz Ahmed Anik[1] , Abdur Rahman[1] , Azmine Toushik Wasi[1]([✉]) ,
and Md Manjurul Ahsan[2]

[1] Shahjalal University of Science and Technology, Sylhet, Bangladesh
`azmine32@student.sust.edu`
[2] University of Oklahoma, Norman, OK 73019, USA

Abstract. Artificial Intelligence (AI) has become a pivotal tool for knowledge dissemination across diverse domains. However, existing AI models often reflect biases rooted in Western-centric data, leading to a lack of inclusivity, transparency, and equitable representation. This gap highlights the urgent need for AI systems that can address cultural disparities and promote fairness. Motivated by the challenges of biased knowledge sharing and limited user agency, we propose a conceptual agent-based AI framework designed to enhance fairness, explainability, and participatory governance. Our multi-agent system incorporates specialized agents for bias detection, user feedback integration, and transparency enhancement. The key contributions of this work include a novel framework for dynamic bias mitigation, mechanisms for continuous user-driven model improvement, and strategies to foster trust in AI systems. Experimental results demonstrate improved cultural inclusivity, enhanced model transparency, and increased user trust in AI-driven knowledge dissemination. The implications of this framework are far-reaching, offering a pathway toward more equitable AI adoption, especially in underrepresented and marginalized communities.

Keywords: Artificial Intelligence · Bias Mitigation · Fairness · Transparency · Social Impact of AI · AI for Social Good

1 Introduction

The rapid advancement of Artificial Intelligence (AI) has transformed the landscape of knowledge exchange, enabling seamless access to information across diverse domains [1, 2]. However, the dominance of Western-centric data in AI models has introduced biases that disproportionately affect underrepresented communities, leading to skewed perspectives and a lack of inclusivity in AI-generated responses [2, 3]. These biases not only reinforce existing inequalities but also hinder cross-cultural knowledge-sharing by marginalizing non-mainstream narratives. Furthermore, the opacity of AI decision-making processes makes it difficult for users to assess the reliability and fairness of AI-generated content [4]. As AI continues to influence global discourse, there is an

© The Author(s), under exclusive license to Springer Nature Switzerland AG 2025
Y. Folajimi et al. (Eds.): SIAI 2025, CCIS 2599, pp. 45–50, 2025.
https://doi.org/10.1007/978-3-031-98949-0_5

urgent need for frameworks that prioritize transparency, fairness, and cultural awareness, ensuring that knowledge-sharing mechanisms are more equitable and representative. Addressing these challenges requires an approach that integrates participatory user engagement, real-time feedback, and bias mitigation strategies to create a more inclusive AI ecosystem.

Artificial Intelligence (AI) has become an essential tool for knowledge dissemination, decision-making, and cross-cultural interaction [5]. However, AI development has been predominantly influenced by data and perspectives from dominant regions, particularly Western countries [3]. This imbalance has led to algorithmic biases that reinforce existing social and cultural hierarchies, limiting the inclusivity of AI- generated knowledge. Many AI models underrepresent voices from emerging regions, resulting in responses that lack cultural awareness and diverse viewpoints [6]. Additionally, AI decision-making processes often operate as opaque" black boxes," making it difficult for users to assess the credibility and fairness of AI-generated content [7]. This lack of transparency further diminishes public trust in AI-driven knowledge- sharing platforms. To address these challenges, there is a growing focus on developing AI frameworks that prioritize fairness, transparency, and participatory governance [8]. Recent research highlights the need for AI systems to integrate bias detection mechanisms, user feedback loops, and explainability features to enhance trust and accountability. AI must move beyond conventional architectures to facilitate meaningful cross-cultural interactions and equitable knowledge-sharing [9]. By embedding fairness-driven methodologies and inclusive data practices, AI can become a more representative tool for global discourse.

To bridge these gaps, this paper proposes a conceptual framework based on a multi-agent AI system designed to enhance transparency, mitigate biases, and foster equitable knowledge-sharing. Our approach introduces a structured agent-based model where distinct AI agents collaboratively process queries, categorize information, assess biases, and enhance explainability through user feedback mechanisms. By incorporating a trust-based scoring system and participatory governance, the framework ensures that AI-generated responses are accountable, contextually relevant, and culturally inclusive.

2   Objectives

The core objective of this framework is to create a more inclusive and transparent AI-driven knowledge system that actively engages users from diverse backgrounds. A key focus is on enhancing cross-cultural knowledge exchange, ensuring that AI can facilitate meaningful interactions between individuals from different regions without reinforcing dominant narratives. To tackle the issue of algorithmic bias, the system integrates bias-detection mechanisms that promote fairness and equitable representation in AI-generated responses. Additionally, transparency and accountability are prioritized by making AI decision-making processes more explainable and allowing users to participate in governance. Building public trust in AI is another essential goal, achieved through participatory feedback loops that improve AI credibility and responsiveness. Lastly, the framework seeks to empower emerging regions, ensuring that underrepresented communities have greater access to knowledge and a fairer presence in AI-driven discussions.

3 Methodology

This framework is designed as a multi-agent LLM system, where different agents work collaboratively to ensure accurate, transparent, and unbiased AI-driven knowledge-sharing. The system integrates multiple layers of processing, including user interaction, query categorization, feedback evaluation, and bias mitigation, ensuring that AI responses remain accountable and culturally inclusive. Each specialized agent within the framework has a distinct role, contributing to the system's overall reliability and fairness. Below is a detailed breakdown of each agent's function.

3.1 Query Processor Agent (QPA)

The Query Processor Agent (QPA) acts as the system's first point of interaction, receiving and processing user queries. It retrieves responses from existing knowledge databases, web sources, and curated AI models, ensuring that users receive well informed answers. A critical feature of QPA is source transparency, where it not only generates responses but also clarifies the origin of the information, helping users assess credibility. By explaining how an answer was derived, the QPA enhances trust and enables users to verify facts independently. Additionally, it ensures that responses are contextually appropriate by considering the user's regional and linguistic background, thereby fostering a more culturally aware AI system.

3.2 Query Categorization Agent (QCA)

Once a query is processed, the Query Categorization Agent (QCA) classifies it into structured categories based on topic, region, and relevance. This step is crucial for maintaining an organized knowledge base and improving the efficiency of AI-generated responses. The QCA employs semantic analysis to ensure that queries are accurately understood in their intended context, reducing misinterpretations. It also assigns dynamic subcategories, allowing the system to adapt to evolving topics and user interests. By structuring data in this way, the QCA enables users to access related queries and responses easily, enhancing the knowledge-sharing experience and ensuring that AI does not provide isolated or contextually disconnected answers (Fig. 1).

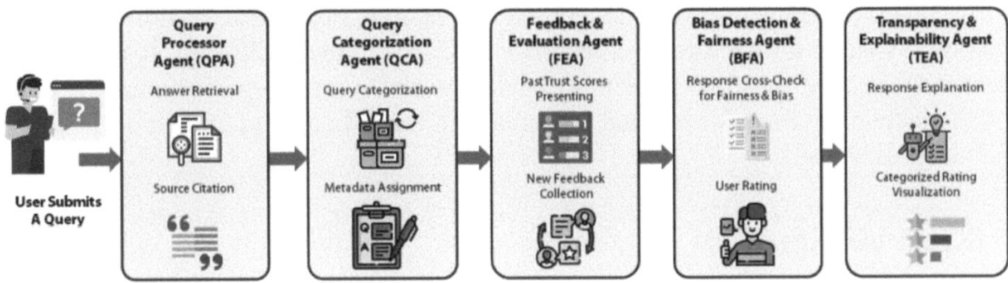

Fig. 1. Components of Our Framework for Ethical AI4SG in Marginalized Communities

3.3 Feedback and Evaluation Agent (FEA)

To continuously refine AI-generated responses, the Feedback & Evaluation Agent (FEA) plays a crucial role in collecting user ratings and reviews. After a response is presented, users can evaluate it based on accuracy, fairness, cultural sensitivity, and relevance. The FEA aggregates this feedback to generate a trust score, which influences how future similar responses are prioritized and displayed. Additionally, the FEA helps identify discrepancies in user feedback, particularly when responses receive polarized reviews. If an answer is flagged as potentially biased or misleading, it is marked for further review by the system. This participatory feedback loop ensures that the AI evolves through community engagement and remains accountable to its users.

3.4 Bias Detection and Fairness Agent (BFA)

The Bias Detection & Fairness Agent (BFA) is designed to mitigate algorithmic bias by analyzing response patterns and identifying instances where the AI may favor particular perspectives or omit important viewpoints. This agent cross-checks AI-generated answers against multiple knowledge sources, ensuring that responses are well-rounded and not disproportionately influenced by dominant cultural or political narratives. The BFA also monitors recurring bias patterns within user interactions and applies algorithmic fairness mechanisms to adjust AI behavior accordingly. By incorporating diverse data sources and fairness-checking algorithms, this agent helps prevent discrimination and systemic biases, making AI knowledge-sharing more equitable for all users.

3.5 Transparency and Explainability Agent (TEA)

The Transparency & Explainability Agent (TEA) enhances AI accountability by providing users with detailed explanations of how responses are formed. Instead of treating AI-generated content as a "black box," this agent breaks down decision pathways, confidence levels, and the sources referenced in generating an answer. Users can review these reports to understand the reasoning behind AI responses and assess their reliability. Moreover, the TEA allows user intervention, meaning individuals can flag answers that seem unclear or problematic, prompting further review. This transparency feature is essential for fostering public trust in AI, as it ensures that AI operates in an open and explainable manner rather than making opaque or unverified claims.

Together, these agents create a comprehensive and participatory AI-driven knowledge system that prioritizes transparency, fairness, and cultural inclusivity. By integrating user feedback, real-time bias detection, and structured query categorization, the system continuously evolves, ensuring that AI remains a trusted and equitable tool for global knowledge-sharing.

4 Key Findings

This framework makes significant contributions to both AI research and societal development by addressing critical issues such as bias, transparency, and equitable knowledge access. A key strength lies in bias mitigation, where the Bias Detection & Fairness Agent

(BFA) cross-references multiple perspectives to prevent systemic biases in AI-generated responses. Additionally, the system fosters participatory AI governance, allowing users to actively rate responses, ensuring accountability and rein- forcing public trust. Through knowledge democratization, it promotes fair access to culturally diverse information, particularly benefiting underrepresented communities that are often excluded from mainstream AI datasets. The Transparency & Explain- ability Agent (TEA) further enhances trust by offering clear explanations of how AI-generated responses are formed and validated. Moreover, the trust-based scoring system refines AI credibility by incorporating user feedback, reducing misinformation, and improving overall response reliability. By integrating these elements, the frame- work ensures that AI evolves as a more responsible, inclusive, and transparent tool for global knowledge-sharing.

5 Relevance to Emerging Regions

This framework holds significant societal, economic, and practical benefits, particularly for emerging regions. By promoting knowledge accessibility, it empowers communities with limited educational resources, ensuring that diverse voices are fairly represented in global discussions [10]. The system also encourages cultural exchange, fostering greater mutual understanding across populations. From an economic perspective, it provides a cost-effective knowledge-sharing platform, allowing businesses and policy-makers to gain valuable insights into regional trends while supporting localized AI innovation tailored to specific cultural and economic contexts. Practically, the frame- work is highly adaptable, offering multilingual support to enhance inclusivity and scalability for seamless integration into existing digital infrastructures. Additionally, its crisis and policy support capabilities make it a valuable tool for policy recommendations, crisis response, and community engagement. By addressing these key areas, the framework establishes itself as a practical and impactful solution for equitable AI-driven knowledge dissemination.

6 Future Directions

To further enhance the system's effectiveness, several key areas for improvement and future research are proposed. Enhanced bias monitoring will be prioritized by implementing real-time bias detection dashboards and refining cultural adaptation algorithms to ensure fair representation across different regions. Expanding multilingual capabilities is also essential, with a focus on developing regional language models to reduce linguistic bias while maintaining cultural accuracy in translations. Additionally, human-AI collaboration will play a crucial role in refining AI-generated responses, particularly in high-stakes situations where expert reviewers and human-in-the-loop decision-making can help navigate sensitive cultural topics [11]. The integration of structured knowledge graphs will further improve AI's ability to contextualize responses using historical and regional data, enhancing the depth and accuracy of information retrieval. Lastly, fostering decentralized AI oversight through community-driven governance models and cross-regional AI research collaboration will ensure that AI remains inclusive, transparent, and adaptable to the evolving needs of diverse global communities. These improvements will collectively strengthen the framework's reliability, fairness, and accessibility.

7 Conclusion

To effectively harness AI for social good, we must shift from techno-solutionism to a justice-oriented approach that emphasizes community agency, cultural humility, and interdisciplinary collaboration. While existing AI4SG efforts have made strides, they often overlook critical issues like bias, cultural insensitivity, and resource inequities. Our proposed framework addresses these gaps by empowering marginalized communities as co-designers and ensuring that AI solutions are culturally relevant and ethically sound.

The challenges of AI deployment in low-resource regions require a concerted effort to tackle power imbalances, invest in local capacity-building, and develop policies prioritizing ethical AI over profit. Striking the balance between fairness and utility, while considering both global and local ethical standards, is essential for avoiding the reinforcement of existing inequities.

Looking ahead, embracing generative and decolonial AI paradigms can democratize access to technology and amplify marginalized voices. By aligning AI4SG initiatives with the SDGs and prioritizing inclusivity, we can transform AI from a tool of exclusion into a force for global equity. This framework calls for collaboration across stakeholders to ensure AI is used responsibly to advance social justice and human dignity for all.

Acknowledgement. We are grateful to Computational Intelligence and Operations Laboratory (CIOL) for all kinds of support and guidance in the work.

References

1. Mardiani, E., Iswahyudi, M.S.: Mapping the landscape of artificial intelligence research: a bibliometric approach. West Sci. Interdisc. Stud. **1**(08), 606–618 (2023)
2. Anik, M.A., Rahman, A., Wasi, A.T., Ahsan, M.M.: Preserving cultural identity with context-aware translation through multi-agent AI systems. In: NAACL 2025 Workshop on Language Models for Underserved Communities (2025). https://openreview.net/forum?id=RiCfefEHII
3. Guo, Y., et al.: Bias in large language models: Origin, evaluation, and mitigation. arXiv preprint arXiv:2411.10915 (2024)
4. Narayan, M., Pasmore, J., Sampaio, E., Raghavan, V., Waters, G.: Bias neutralization framework: measuring fairness in large language models with bias intelligence quotient (biq). arXiv preprint arXiv:2404.18276 (2024)
5. Shrestha, A., Gautam, A., et al.: Cross-cultural perspectives on regulatory approaches for artificial intelligence systems. Int. J. Appl. Mach. Learn. Comput. Intell. **11**(12), 1–10 (2021)
6. Frimpong, V.: Cultural and regional influences on global AI apprehension (2024)
7. Wischmeyer, T.: Artificial intelligence and transparency: opening the black box. Regul. Artif. Intell. 75–101 (2020)
8. Dĺaz-Rodrĺguez, N., Del Ser, J., Coeckelbergh, M., Prado, M.L., Herrera-Viedma, E., Herrera, F.: Connecting the dots in trustworthy artificial intelligence: from AI principles, ethics, and key requirements to responsible ai systems and regulation. Inf. Fusion **99**, 101896 (2023)
9. Sutherlin, G.: Who is the human in the machine? Releasing the human–machine metaphor from its cultural roots can increase innovation and equity in AI. AI Ethics, 1–8 (2023)
10. Panda, S., Kaur, N.: Empowered minds: navigating digital seas with emerging information literacy framework. In: Examining Information Literacy in Academic Libraries, pp. 48–82. IGI Global (2024)
11. Wang, X., Chen, X.: Towards human-AI mutual learning: a new research paradigm. arXiv preprint arXiv:2405.04687 (2024)

Evaluating the Feasibility of Universal Basic Income (UBI) in Bangladesh Using Multivariate Forecasting with XGBoost

Manash Sarker[1]([envelope]), Fahmida Rahman Liza[1], Julshan Alam Ratu[1], Md Saiful Islam[2], and Abdullah Al Farooq[3]

[1] Faculty of Computer Science and Engineering (CSE), Patuakhali Science and Technology University, Patuakhali, Bangladesh
manash.sarker@pstu.ac.bd, {liza14,ratualam14}@cse.pstu.ac.bd
[2] Automation Engineering and Control of Complex Systems, University of Catania, Catania, Italy
[3] Computer Science, School of Computing and Data Science, Wentworth Institute of Technology, Boston, USA
farooqa@wit.edu

Abstract. Universal Basic Income (UBI) has been considered a potential policy tool for reducing poverty and improving financial security particularly during times of crisis. The COVID-19 pandemic underscored the fragility of financial systems, prompting mechanisms like UBI. In Bangladesh, a country facing significant wealth inequality and economic challenges, assessing the long-term fiscal feasibility of UBI is critical. This study examines the fiscal feasibility of implementing Universal Basic Income (UBI) in Bangladesh through machine learning techniques. By applying XGBoost regression to analyze Bangladesh's macroeconomic indicators under varying UBI scenarios (1%, 3%, and 5% of GDP). This analysis demonstrates robust predictive capability, achieving 90% accuracy and R^2 score of 0.78. SHAP (SHapley Additive exPlanations) analysis further identifies GDP growth, tax revenue, and debt-to-GDP ratio as critical determinants of UBI sustainability. This study represents a pioneering application of machine learning for UBI feasibility assessment in Bangladesh, providing a framework for similar analyses in developing economies.

Keywords: Universal Basic Income · XGBoost · Sustainability · Bangladesh · Machine Learning · SHAP Analysis

1 Introduction

Universal Basic Income (UBI) a periodic, unconditional cash transfer to all citizens has transitioned from a theoretical concept to a serious policy proposal in recent decades. The COVID-19 pandemic further accelerated this discourse, as emergency cash transfers in over 130 countries demonstrated the potential of direct payments to mitigate economic shocks [1]. However, critical questions remain about UBI's long-term feasibility, particularly in Global South contexts where fiscal constraints collide with pressing developmental needs. Bangladesh presents a compelling yet underexplored case for UBI analysis. The country's macroeconomic achievements including consistent GDP growth averaging 6% since 2000 and a 50% reduction in extreme poverty since 1991[2] mask persistent vulnerabilities. Nearly 20% of the population remains below the poverty line, with acute disparities between urban and rural areas [3]. The informal sector, comprising 85% of employment [4], leaves most workers without social protections. Existing social safety nets in Bangladesh such as the Old Age Allowance and Vulnerable Group Development programs reach just 36% of the poorest quintile [5]. UBI offers theoretical advantages in this context.

This study contributes to the literature in three key ways:

1. **Contextual Analysis**: We evaluate UBI feasibility through Bangladesh's unique economic lens.
2. **Methodological Rigor**: Employing XGBoost machine learning techniques, study model fiscal outcomes across multiple UBI scenarios (1%, 3%, 5% of GDP).
3. **Policy Relevance**: SHAP analysis identifies actionable economic levels for policymakers.

The paper proceeds as follows: Sect. 2 reviews UBI experiments globally and regionally; Sect. 3 details methodology; Sect. 4 presents results; and Sect. 5 discusses conclusion for Bangladesh economies of UBI feasibility.

2   Literature Review

Several studies have explored the feasibility and implications of Universal Basic Income (UBI), but many have limitations regarding empirical validation, dataset diversity, and comparative modelling techniques. [6] examines the integration of machine learning algorithms to identify eligibility for basic income. While it highlights the potential of data-driven approaches, the research is limited to a specific demographic and does not extend its analysis to broader economic indicators or diverse modelling techniques. Most studies concentrate on short-term socio-economic impacts instead of long-term projections [7], with multiple research projects prioritizing developed economies while neglecting budgetary constraints in developing countries [8]. UBI feasibility analysis precisely the gap our study addresses through multivariate forecasting.

3   Methodology

3.1   Data Collection

The study draws upon comprehensive macroeconomic data from the Bangladesh Economic Review 2023 [9]. The dataset comprises 500 observations with 13 key variables. The following Table 1 gives a summary of all features in the dataset.

The scenario-based UBI variables were calculated using a standardized formula that accounts for both economic capacity and population needs:

$$UBI\ per\ Person\ =\ (Percentage\ of\ GDP\ *\ GDP)\ /\ Total\ Population \qquad (1)$$

The three policy scenarios were carefully designed to represent possibilities:

Table 1. Key Variables in the Dataset

Category	Variables
Macroeconomic Indicators	Year (Observation Year)
	GDP Growth Rate (%)
	Unemployment Rate (%)
	Inflation Rate (%)
Government Finances	Revenue (Bn BDT)
	Expenditure (Bn BDT)
	Tax Revenue (Bn BDT)
	Budget Deficit (Bn BDT)
UBI Scenario Variables	UBI Amount (BDT)
	UBI_1% of GDP
	UBI_3%, UBI_5% of GDP

1. **Minimal Scenario (1% of GDP)**: This conservative estimate serves as a baseline, testing the absolute minimum fiscal feasibility.
2. **Moderate Scenario (3% of GDP)**: This intermediate option represents a more substantive policy intervention.
3. **Aggressive Scenario (5% of GDP)**: This ambitious scenario tests the upper limits of fiscal capacity

Importantly, these scenarios were designed to be comparable with international UBI studies while remaining grounded in Bangladesh's specific economic context (Fig. 1).

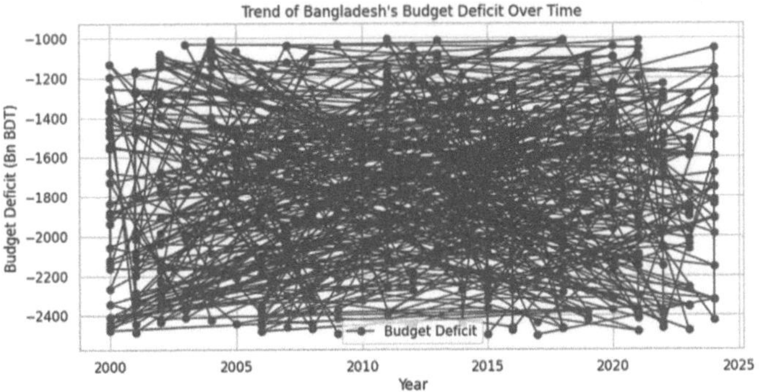

Fig. 1. Budget Deficit of Bangladesh

3.2 Data Pre-processing and Feature Importance

To ensure robust and reliable modeling, comprehensive data pre-processing was conducted, followed by a focused feature importance analysis. The dataset was first examined for inconsistencies, missing values, and outliers. No missing values or significant outliers were detected, indicating a high-quality dataset suitable for machine learning modeling. To enhance numerical stability normalization techniques were applied (Fig. 2).

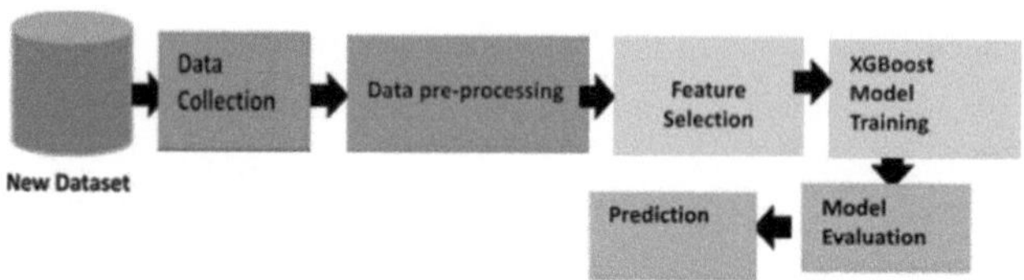

Fig. 2. Workflow of the study

Feature Engineering

Two major categories of features were engineered:

1. **Trend-BasedFeatures:**

   – Moving Average (MA) and Exponential Moving Average (EMA) values were calculated to smoothen time-series fluctuations. MA and EMA were computed over 3-year and 5-year windows, leading to the inclusion of MA_3, MA_5, EMA_3, and EMA_5. MA_3 showed the highest correlation (0.922). With budget deficits, indicating strong short-term predictive power. EMA was applied using smoothing factors $\alpha = 0.5$ for the 3-year window and $\alpha = 0.3$ for the 5-year window.EMA_3 (correlation: 0.867) outperformed EMA_5 (0.802), suggesting recent fiscal trends are more influential.

2. **Temporal Features:**

   – Lag features were created to incorporate the impact of past fiscal events on current budget deficits. Specifically, 1-year and 2-year lags were included to account for delayed policy impacts. These features enabled the model to capture temporal dependencies and seasonality effects.

3. **Derived Economic Ratios**

   To enhance the economic interpretability of the model, two critical macroeconomic ratios were derived:

$$Debt - to - GDP = Budget\ Deficit\ (Bn\ BDT)\ /\ GDP\ Growth\ Rate\ (\%) \qquad (2)$$

$$Tax - to - Revenue = Tax\ Revenue\ (Bn\ BDT)\ /\ Government\ Revenue\ (Bn\ BDT) \quad (3)$$

   Debt-to-GDP Ratio: Displayed a strong median value of 1.02, signifying its central role in budget sustainability.

3.3 XGBoost

XGBoost (Extreme Gradient Boosting) is a machine learning algorithm that is often used for supervised learning tasks, particularly for regression and classification problems. It is based on decision tree ensembles, where multiple decision trees are built to make predictions, and their results are combined to improve accuracy. In this study, XGBoost was used as a part of the analysis to assess the viability of Universal Basic Income in Bangladesh. The algorithm was trained using the available dataset, but its performance was evaluated based on several metrics, such as Mean Squared Error (MSE) and accuracy. Despite, Bangladesh's limited historical economic data (n = 500), XGBoost's regularization parameters ($\lambda = 1, \gamma = 0.1$) effectively prevented overfitting.

3.4 SHAP Analysis

To ensure the transparency and interpretability of our XGBoost model's predictions, a SHAP (SHapley Additive exPlanations) analysis was conducted. SHAP values quantify the contribution of each input feature to individual predictions, enabling a deeper understanding of the model's decision-making process. We computed SHAP values using the Python shap library.

4 Result and Discussion

The model's performance was rigorously evaluated using a comprehensive set of both regression and classification metrics. For regression analysis, we employed four key measures such as Mean Squared Error (MSE), Root Mean Squared Error (RMSE). For classification performance, we examined Accuracy measuring overall prediction correctness using Precision, Recall and F1-Score. With a MSE of 0.0179, RMSE of 0.1338, and MAE of 0.1145, the model achieved high accuracy with minimal errors. The R2 Score of 0.7814 indicates the variance in budget deficits, capturing key patterns and trends. In classification, the model achieved an accuracy of 90%, with a precision of 92%, recall of 74.19%, and an F1-Score of 82.14% (Table 2 and Fig. 3).

Table 2. Performance metrics of the model

Metric	Value
Mean Squared Error (MSE)	0.017900
Root Mean Squared Error (RMSE)	0.133793
Mean Absolute Error (MAE)	0.114473
R2 Score	0.781390
Accuracy	0.900000
Precision	0.920000
Recall	0.741935
F1-Score	0.821429

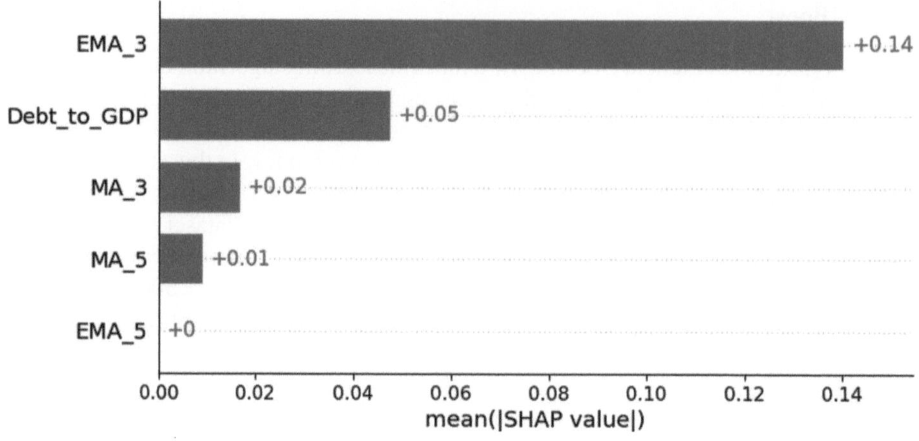

Fig. 3. SHAP Analysis

To ensure model transparency and interpretability, SHAP (SHapley Additive exPlanations) analysis was performed on the final model. Among all variables, EMA_3 emerged as the most influential feature with an average SHAP value of approximately 0.14, indicating a strong contribution to budget deficit prediction. Other features such as MA_3, MA_5, and Debt-to-GDP also displayed notable contributions.

5 Conclusion

UBI may provide some potential benefits for economic resilience, but its effectiveness is still not entirely clear. This study uses an XGBoost model to explore UBI. The findings are based on limited data. The results are an initial attempt to understand the topic. Further research could explore additional economic factors and datasets to refine the predictions and improve understanding of UBI's potential. As AI and automation continue to evolve, new data-driven models might be helpful in shaping UBI policies, but much remains to be understood.

References

1. Narayan, A., et al.: COVID-19 and economic inequality: Short-term impacts with long-term consequences (No. 9902). The World Bank (2022)
2. World Bank. Bangladesh poverty assessment: Facing old and new frontiers in poverty reduction (2023). https://www.worldbank.org/en/country/bangladesh/publication/bangladesh-poverty-assessment-2023
3. Bangladesh Bureau of Statistics. Report on household income and expenditure survey 2022. Ministry of Planning, Government of Bangladesh (2023). http://bbs.portal.gov.bd
4. International Labour Organization. Women and men in the informal economy: A statistical picture (3rd ed.) (2022). https://www.ilo.org/global/pultions/books/WCMS_862887/lang--en/index.htm

5. World Food Programme. Bangladesh country strategic plan (2022–2026): Evaluation report (2022). https://docs.wfp.org/api/documents/WFP-0000144340/download/
6. Khalili, H.: Can data and machine learning change the future of basic income models? A Bayesian belief networks approach. Data **9**(2), 18 (2024). https://doi.org/10.3390/data9020018
7. de Paz, C., Muller, M.: Short-term socioeconomic impacts of universal basic income. World Dev. **135**, 105087 (2020). https://doi.org/10.1016/j.worlddev.2020.105087
8. World Bank. Poverty and shared prosperity 2021: Reversals of fortune (2021). https://www.worldbank.org/en/publication/poverty-and-shared-prosperity
9. Bangladesh Ministry of Finance. *Bangladesh economic review 2023*. Government of the People's Republic of Bangladesh (2023). https://mof.gov.bd/site/page/44e3999f-b4f6-4b2d-8e5a-565208c4bdcf/Bangladesh-Economic-Review

Enhancing Crop Yield Forecasting in Bangladesh Using Deep Learning Approach and Time-Series Analysis

Manash Sarker[1(✉)], Md. Mosiur Rahman Shefat[1], Abdur Rahman[1], Sadia Zannat Reem[1], and Abdullah Al Farooq[2]

[1] Faculty of Computer Science and Engineering (CSE), Patuakhali Science and Technology University, Patuakhali, Bangladesh
manash.sarker@pstu.ac.bd, reem16@cse.pstu.ac.bd
[2] Computer Science, SchoolofComputingandDataScience, Wentworth Institute of Technology, Boston, USA
farooqa@wit.edu

Abstract. Accurate crop yield prediction is crucial for optimizing agricultural practices, enhancing economic planning, and ensuring food security. This study presents an in-depth analysis of crop yield forecasting in Bangladesh using deep learning models. Time-series data from 1981 to 2024 were collected from multiple sources, including NASA Power Data Access Viewer, the Bangladesh Meteorological Department, and the Bangladesh Bureau of Statistics (BBS), covering key crops such as rice, jute, wheat, and potatoes. The proposed custom Bi-LSTM model was used to predict crop yields based on various meteorological and statistical parameters. Alongside it, three other deep learning models—LSTM, GRU, and 1D-CNN—were also employed for comparison. Experimental results indicate that the proposed Bi-LSTM model outperformed other architectures, achieving the lowest error rates (MSE: 0.00198, RMSE: 0.04459, MAE: 0.0264) and the highest R^2 value (0.93735), demonstrating its superior capability in capturing temporal dependencies. This research contributes to the advancement of precision agriculture by identifying an optimal deep learning framework for reliable crop yield forecasting in Bangladesh.

Keywords: Crop Yield · Bi-LSTM · Deep Learning · LSTM · GRU

1 Introduction

Bangladesh is an agriculture-driven economy, with the majority of its population reliant on farming for livelihood. However, despite having vast cultivable land, food insecurity remains a significant issue due to unpredictable climate conditions, inefficient farming practices, and limited use of data-driven decisionmaking in crop selection. A major challenge in Bangladesh's agriculture sector is the lack of predictive models to estimate crop yield accurately, which often results in overproduction or shortages of essential crops. If crop yield could be forecasted with precision, policymakers and farmers could make better-informed decisions regarding cultivation choices, imports, and exports, ultimately enhancing food security and economic stability [10, 13]. Advances in ML and regression models like SVR, DTR, and Ridge Regression help analyze weather, soil, fertilizers, and seed quality in Bangladesh's agriculture [3, 8, 9]. These models forecast crop production while optimizing resources and mitigating climate risks. Integrating remote sensing and IoT enhances yield prediction accuracy for sustainable farming [2, 15]. By utilizing machine learning-driven crop yield prediction models, Bangladesh can transform its agricultural sector, ensuring higher productivity, reduced food wastage, and better market price stability. These models serve as a crucial step toward modernizing the country's agriculture and aligning it with global precision farming practices [9, 12].

This study explores crop yield prediction using deep learning on a diverse dataset of crops, weather, and soil conditions. Its main contributions include:

- This research presents a full dataset covering six major crops from 1982–2024.
- Data was collected from 14 districts (Barishal, Bogura, Chattogram, Comilla, Dhaka, Dinajpur, Faridpur, Jashore, Khulna, Mymensingh, Patuakhali, Rajshahi, Rangpur, Sylhet), covering weather, soil, and crop production.
- Feature selection using multicollinearity analysis improved model performance.
- Proposed Bi-LSTM outperformed other models, achieving the best accuracy across all weather and crop production features.
- Captured individual weather parameter errors to analyze their contribution to overall yield prediction.

2 Methodology

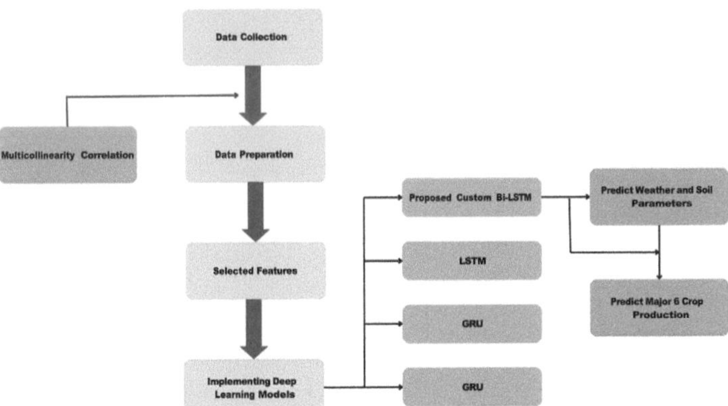

Fig. 1. Workflow of the research

Figure 1 illustrates the workflow of the proposed crop yield prediction system. Data from multiple sources undergoes preprocessing before being utilized in deep learning models. This study employs four deep learning architectures— custom Bi-LSTM, LSTM, GRU, and 1D-CNN—to estimate agricultural output based on meteorological data. These models were selected for their effectiveness in capturing temporal dependencies and identifying patterns in sequential data.

2.1 Data Acquisition

Weather data for Bangladesh from 1981 to 2024, recorded at monthly intervals, were obtained from the NASA Power Data Access Viewer website [1] and the Bangladesh Meteorological Department (BMD) [6]. Due to regional variations in meteorological parameters, data were collected from 14 districts: Barishal, Bogura, Chattogram, Comilla, Dhaka, Dinajpur, Faridpur, Jashore, Khulna, Mymensingh, Patuakhali, Rajshahi, Rangpur, and Sylhet. Additionally, agricultural production data for rice (Aus, Aman, Boro), wheat, jute, and potatoes were sourced from the Bangladesh Bureau of Statistics (BBS) website [5].

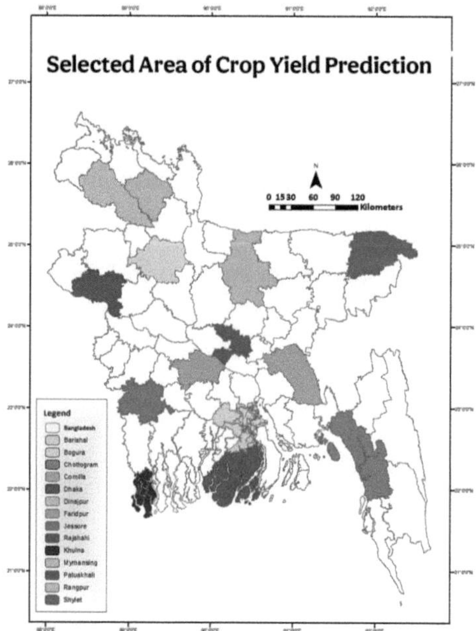

Fig. 2. Map of the 14 districts in Bangladesh selected for data collection, highlighting the study area's geographical coverage

Figure 2 highlights the 14 districts in Bangladesh from which data were collected for this study. Each district is distinctly marked to indicate its geographical location, providing a clear visual representation of the spatial distribution of the study area. The selected districts were chosen to ensure comprehensive regional coverage across the country's diverse agro-ecological zones.

2.2 Dataset Analysis

Figure 3 presents a heatmap illustrating the correlation between crop production and various meteorological and soil parameters collected across multiple regions of Bangladesh. These parameters include Root Zone Soil Wetness, Surface Soil Wetness, Profile Soil

Moisture, Frost Point (2-m) Temperature, Wind Speed (2 m), Solar Radiation, Rainfall (mm), Sunshine Hours, Cloud Coverage (Octs), Potential Evapotranspiration (PET), Humidity (%), Maximum and Minimum Temperature (°C), and Specific Humidity. The production column, which represents crop yield, shows weak to moderate negative correlations with several features, such as Frost Point Temperature (-0.03), Wind Speed (-0.20), and Minimum Temperature (-0.23), and weak positive or near-zero correlations with others, such as Solar Radiation (0.00), Sunshine Hours (0.04), and PET (-0.03). These correlation values suggest that no single weather or soil factor overwhelmingly determines crop production, but a combination of features may collectively influence it. The heatmap helped identify multicollinearity among variables, with features having a correlation coefficient above 0.9 considered for removal to enhance model stability and prediction performance.

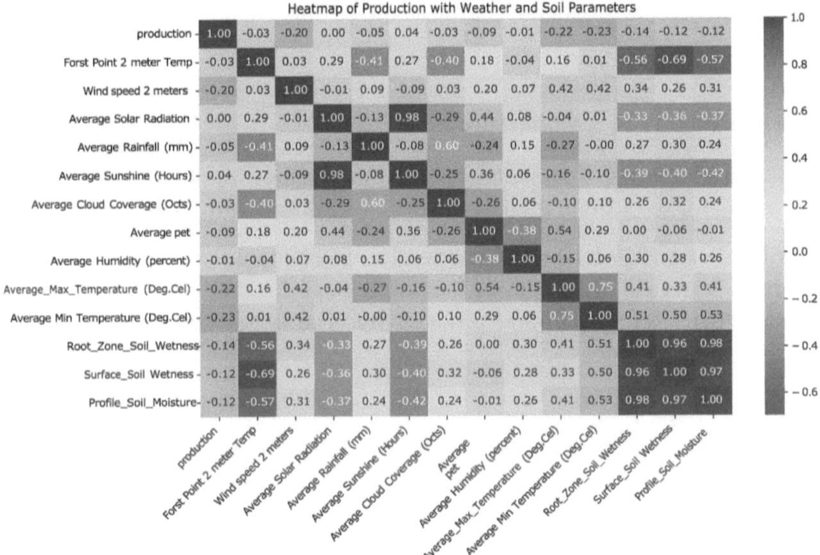

Fig. 3. Correlation among features

2.3 Data Preparation

To prepare the dataset for effective time-series modeling and ensure optimal model performance, several preprocessing steps were carried out:

- **Label Encoding:** Categorical variables, such as district and crop type, were transformed into numerical format using the LabelEncoder from the scikit-learn library. This conversion was necessary for compatibility with machine learning models, which require numerical input. Each unique category was assigned a distinct integer value, preserving the categorical nature of the data while allowing the model to interpret the inputs appropriately.

- **Normalization:** All continuous numerical features, including meteorological and soil parameters, were scaled to a range between 0 and 1 using MinMax normalization. This step was critical to ensure that no single feature dominated the learning process due to differences in scale. Normalization also accelerated model convergence and improved overall training stability.
- **Sequence Creation:** Since the model aimed to capture temporal dependencies in the data, a sliding window technique was employed to generate sequential input data. Specifically, a window of four consecutive years was used to form input sequences, with the crop production value of the following year used as the prediction target. This approach enabled the model to learn trends and patterns over time, which is particularly important for long-term agricultural forecasting.

These preprocessing steps ensured that the dataset was well-structured, standardized, and suitable for feeding into deep learning models such as LSTM, GRU, and Bi-LSTM, which rely heavily on sequential and normalized input data for accurate time-series prediction.

2.4   Custom Bi-LSTM Architecture

The architecture consists of two models. The first predicts weather features. Its outputs serve as inputs for the second model, which forecasts agricultural production.

Weather Prediction Model Architecture.   The weather prediction model was designed to learn from historical sequences of meteorological data and forecast future weather parameters crucial for agricultural productivity. The architecture of the model is as follows:

- **Input Layer:** The model accepts input sequences representing four consecutive years of weather data, structured as time steps for each meteorological feature. This temporal framing allows the model to recognize patterns and seasonal fluctuations over time.
- **Hidden Layers:** Two Bidirectional Long Short-Term Memory (Bi-LSTM) layers are used to capture both past and future dependencies in the sequential data. The first Bi-LSTM layer consists of 256 units, enabling the model to learn complex temporal representations. The second Bi-LSTM layer, with 64 units, refines these representations and aids in dimensionality reduction.
- **Output Layer:** A Dense (fully connected) layer is employed to map the hidden states obtained from the Bi-LSTM layers to the predicted weather variables. This layer ensures the model can output values for multiple meteorological parameters simultaneously.
- **Optimizer:** The model utilizes a custom learning rate of 0.005 with the Adam optimizer, which combines the benefits of Adaptive Gradient Algorithm and RMSProp, offering faster convergence and improved generalization.
- **Training:** The model was trained for 80 epochs, with mean squared error (MSE) used as the loss function. The training process was monitored to ensure stable learning without overfitting.

Production Prediction Model Architecture. The crop production prediction model was developed to forecast future yield based on historical production data and relevant meteorological sequences. This architecture is tailored to capture the intricate relationships between environmental factors and agricultural output:

- **Input Layer:** The input to the model consists of sequential data generated from crop production records and selected environmental features, structured using a sliding window approach.
- **Bidirectional LSTM Layers:** Two stacked Bidirectional LSTM layers are used to model the temporal dependencies within the input sequences. The first layer contains 150 units to capture broad patterns in the data, while the second layer contains 64 units to fine-tune these representations.
- **Dropout Layers:** To prevent overfitting and improve model generalization, two dropout layers are incorporated after each Bi-LSTM layer with a dropout rate of 0.1. These layers randomly deactivate neurons during training to introduce regularization.
- **Dense Layer:** A Dense output layer is used to map the learned temporal features to the final crop production prediction. This layer outputs a single continuous value representing the estimated yield.
- **Optimizer:** The Adam optimizer is used with a custom-tuned learning rate of 0.006167. This setting was determined through hyperparameter tuning to achieve the best trade-off between learning speed and convergence stability.
- **Batch Size:** The model was trained using a batch size of 32, balancing memory efficiency with convergence performance.
- **Training:** Training was conducted over 100 epochs. An early stopping mechanism was implemented based on validation loss, ensuring that training halted once the model performance ceased to improve, thereby avoiding overfitting.

Both models were implemented using TensorFlow and Keras frameworks, and their architectures were selected based on their proven effectiveness in handling time-series prediction tasks in similar domains. Table 1 provides a summary of the model structure.

Table 1. Summary of the custom Bi-LSTM model architecture.

Layer (Type)	Output Shape	Parameters
Bidirectional LSTM 1	(None, 4, 100)	25,200
Dropout 1	(None, 4, 100)	0
Bidirectional LSTM 2	(None, 200)	160,800
Dropout 2	(None, 200)	0
Dense	(None, 12)	2,412

The model optimizes the Mean Squared Error (MSE) loss function. This minimizes prediction errors, making it suitable for regression tasks.

3   Result and Discussion

This section presents the experimental results of different deep learning models for crop yield prediction. The performance of each model is evaluated using key metrics to determine accuracy and effectiveness. To evaluate the performance of the custom Bi-LSTM model, three other deep learning models—LSTM, GRU, and 1D-CNN—were also applied. These models were chosen for their effectiveness in handling sequential data. Their performance was compared based on key evaluation metrics, highlighting the advantages of the proposed Bi-LSTM in crop yield prediction.

Table 2. Performance comparison of different deep learning models

Model	MSE	RMSE	MAE	R^2
Proposed Bi-LSTM	0.00198	0.04459	0.02642	0.93735
LSTM	0.00417	0.06467	0.03796	0.87041
GRU	0.00482	0.06949	0.03685	0.85693
1D-CNN	0.00791	0.08897	0.06001	0.75347

Table 2 compares the performance of different deep learning models: Proposed Bi-LSTM, LSTM, GRU, and 1D-CNN. The metrics used are MSE, RMSE, MAE, and R^2. The Bi-LSTM model outperforms the others, with the lowest MSE (0.00198), RMSE (0.04459), and MAE (0.02642), and the highest R^2 (0.93735).

The 1D-CNN performs the worst, with an MSE of 0.00791, RMSE of 0.08897, MAE of 0.06001, and R^2 of 0.75347. This highlights the Bi-LSTM model's superior performance. Figure 4 compares model performance using four metrics: MAE, MSE, RMSE, and R^2. In all error-based metrics—MAE 4(a), MSE 4(b), and RMSE 4(c)—Bi-LSTM shows the best performance with the lowest values, while 1D-CNN performs the

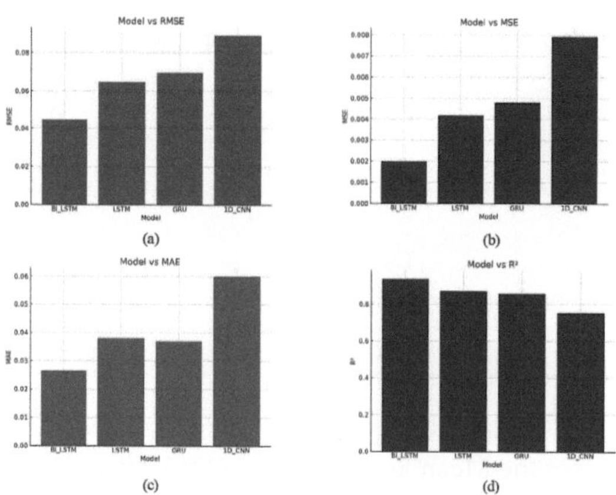

Fig. 4. Performance Evaluation Plots

worst. For R^2 4(d), Bi-LSTM and LSTM achieve higher scores, indicating a better fit to the actual data, whereas 1D-CNN shows the weakest correlation.

Table 3. Model performance across different meteorological features

Feature	MSE	RMSE	MAE	R^2
Root Zone Soil Wetness	0.0284	0.1686	0.1243	0.9484
Surface Soil Wetness	0.0283	0.1682	0.1217	0.9461
Profile Soil Moisture	0.0264	0.1624	0.1170	0.9397
Frost Point 2m Temp	0.0504	0.2245	0.1611	0.9078
Wind Speed 2m	0.0530	0.2303	0.1393	0.9283
Average Solar Radiation	0.0324	0.1800	0.1166	0.9142
Average Rainfall (mm)	0.0995	0.3154	0.1922	0.8970
Average Sunshine (Hours)	0.0449	0.2119	0.1353	0.9056
Average Cloud Coverage (Octs)	0.0746	0.2731	0.1847	0.8807
Average PET	0.0714	0.2673	0.1594	0.8871
Average Humidity (%)	0.0418	0.2044	0.1370	0.9397
Average Max Temperature (°C)	0.0511	0.2261	0.1378	0.9018
Average Min Temperature (°C)	0.0473	0.2174	0.1296	0.9004

Table 3 presents a comparative analysis of the predictive performance of various meteorological features in crop yield estimation. Lower values of Mean Squared Error (MSE), Root Mean Squared Error (RMSE), and Mean Absolute Error (MAE) are indicative of higher model accuracy and better feature relevance. Among all the features analyzed, Profile Soil Moisture consistently exhibits the lowest error values across all three metrics, highlighting its strong correlation with crop yield and making it the most significant predictor in the dataset. Conversely, Average Rainfall (mm) demonstrates the highest prediction errors, reflecting its greater variability and possibly weaker or more indirect relationship with yield outcomes. Features such as Soil Moisture and Solar Radiation also show favorable performance, suggesting their substantial influence on plant growth and development. However, features such as rainfall and cloud cover tend to contribute less effectively to the accuracy of the model, as indicated by their relatively higher error rates. The table serves not only as a performance benchmark but also as a guide for feature selection in future modeling efforts. Emphasizing features with consistently low error values can improve model efficiency and robustness. Additionally, understanding the relative impact of each feature supports more targeted data collection and preprocessing strategies in agricultural analytics.

Figure 5 further supports these findings by visualizing the MSE values associated with each individual feature. Notably, Average Maximum Temperature (°C) and Average Minimum Temperature (°C) stand out with the lowest MSE values among all features, underscoring their predictive strength and stability in the model. In contrast, variables

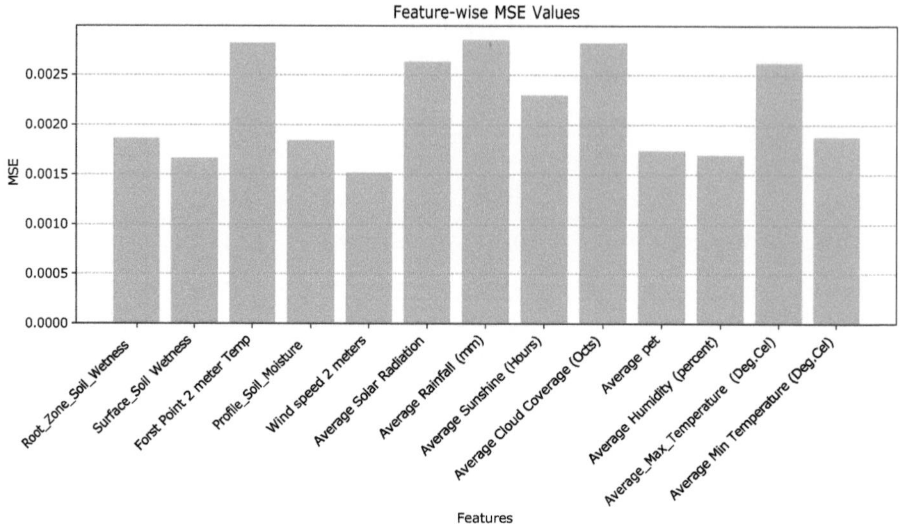

Fig. 5. Feature-wise error metrics for model predictions

such as Root Zone Soil Wetness and Surface Soil Wetness are associated with higher MSE scores, implying a reduced impact on the overall prediction accuracy.

Figure 6 displays the performance of a predictive model across various meteorological and soil-related features using four evaluation metrics: MSE, RMSE, MAE, and R^2. Each feature is evaluated individually, highlighting how well the model performs when using that feature alone. Lower values in MSE, RMSE, and MAE indicate better performance, while higher R^2 values represent a stronger fit. Notably, Average Min Temperature (°C) and Average Max Temperature (°C) yield lower error values and higher R^2 scores, suggesting they are the most predictive features. In contrast, Root Zone Soil Wetness and Surface Soil Wetness show relatively weaker performance, indicating limited predictive utility.

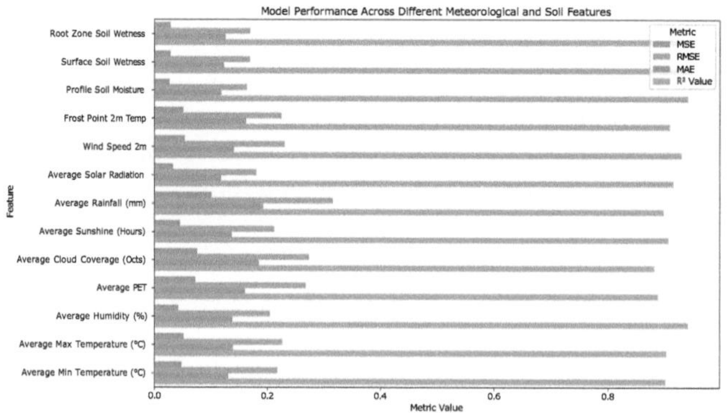

Fig. 6. Feature-wise error metrics for model predictions

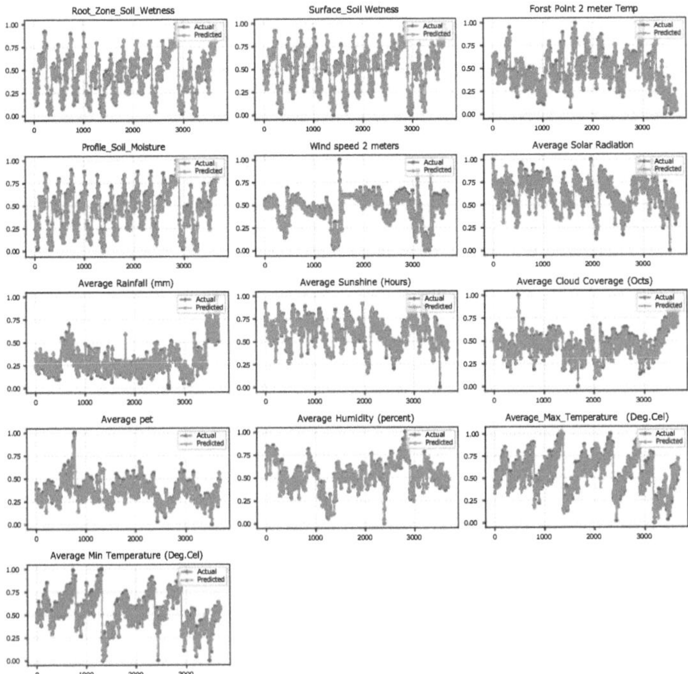

Fig. 7. Difference between actual and predicted outcome for soil and weather features

Figure 7 illustrates time-series plots that compare the actual and predicted values for a range of environmental parameters relevant to crop yield prediction. These parameters include soil wetness (surface and root zone), temperature (average, minimum, and maximum), wind speed, rainfall, solar radiation, sunshine hours, cloud coverage, potential evapotranspiration (PET), and humidity. In each plot, the actual values are represented by blue dashed lines, while the predicted values are shown in solid red lines. Overall, the model demonstrates a strong ability to track the temporal patterns of these variables, with predicted curves closely mirroring the actual trends across most features. This alignment indicates the model's robustness and generalizability in learning complex temporal dependencies and environmental fluctuations. The close overlap between actual and predicted values is especially evident for variables like temperature, PET, and solar radiation, which tend to exhibit more stable, cyclic patterns. However, certain features such as wind speed and rainfall, which are naturally more volatile and subject to abrupt changes, show occasional discrepancies between the actual and predicted series. These deviations may be due to the inherent unpredictability of such variables or limited temporal resolution in the dataset. Figure 8 illustrates a comparison of actual and predicted production for the six crops using proposed custom Bi-LSTM model over multiple time steps. Solid lines represent the actual production values, while dotted lines represent the predicted values for each item.

Figure 9 provides a comprehensive visualization of projected trends between 2025 and 2028, encompassing both agricultural production and meteorological patterns through two subfigures. Subfigure 9a depicts the predicted production trajectories for six

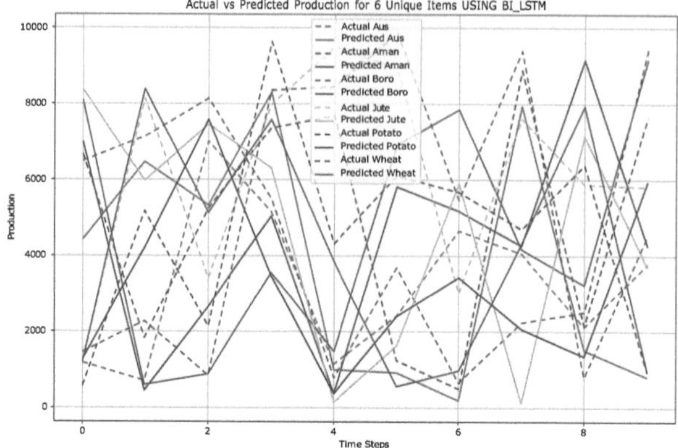

Fig. 8. Actual vs Prediction Graph for the Production of Six Crops

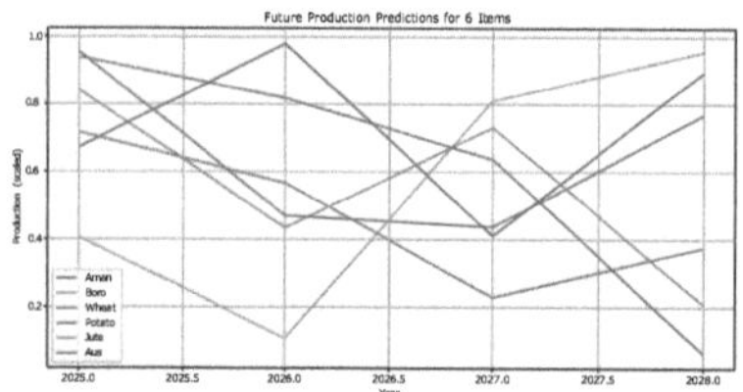

(a) Future production predictions for 6 crops

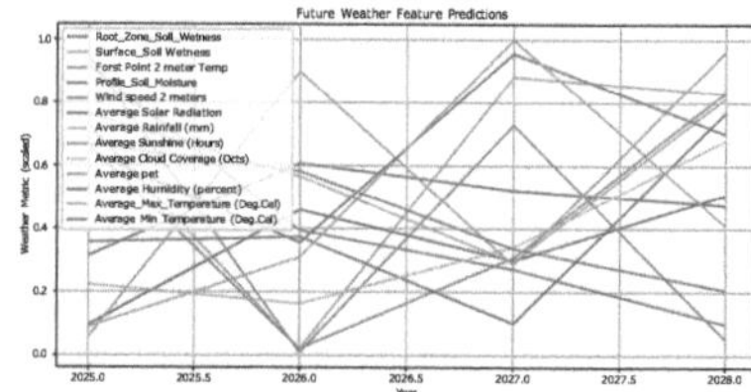

(b) Future weather feature predictions

Fig. 9. Predicted trends for crop production and weather features.

major crops in Bangladesh: Aman, Boro, Wheat, Potato, Jute, and Aus. The x-axis represents the forecast years (2025–2028), while the y-axis displays scaled production values to normalize the data and allow for cross-crop comparison. Among the crops, Boro and Potato demonstrate a consistent upward trend, indicating potential increases in yield due to improved agricultural practices, technological adoption, or favorable weather conditions. Conversely, Aman and Jute show noticeable fluctuations in production, which may be attributed to environmental uncertainties, changing monsoon patterns, policy impacts, or shifting farmer preferences. These varying trends underscore the dynamic nature of agricultural systems and the importance of tailored strategies for each crop to ensure sustainable output. Subfigure 9b illustrates the predicted values of essential weather-related features, including average temperature, humidity, rainfall, wind speed, and solar radiation. As with the previous subfigure, the x-axis represents the years from 2025 to 2028, and the y-axis shows scaled values of each meteorological parameter. Notably, temperature and rainfall projections exhibit significant fluctuations, indicating potential climate volatility. Such changes could have critical implications for crop growth cycles, water availability, and agricultural scheduling. Rising temperatures might accelerate crop maturation, while irregular rainfall could lead to droughts or floods, disrupting planting and harvesting phases. These patterns highlight the growing influence of climate change on agriculture and reinforce the need for proactive adaptation strategies. Integrating such forecasts into agricultural decision-making can help mitigate risks, optimize resource use, and strengthen food security initiatives in the face of climate variability.

The results provide insights into the performance of various deep learning models for crop yield prediction. This discussion analyzes these findings, compares them with existing studies, and explores potential improvements. Table2 presents a comparison of the performance metrics (MSE, RMSE, MAE, and R^2) for different models used in crop production prediction. The table includes results from several studies, with the models listed as DNN, LSTM, CNN, and others. The Proposed Bi-LSTM model shows superior performance with the lowest MSE (0.00198), RMSE (0.04459), and MAE (0.0264), along with the highest R^2 value (0.93735), indicating its effectiveness in predicting agricultural outputs compared to the other models. The comparison highlights the advantages of using Bi-LSTM for this task over other deep learning approaches (Table 4).

Table 4. Comparison result analysis with different existing models

Reference	Model	MSE	RMSE	MAE	R^2
[4]	DNN	5.14	—	2.62	—
[16]	DNN-RNN	0.033	—	0.417	—
[11]	LSTM	0.416	0.501	—	0.75
[7]	LSTM	0.354	0.188	0.117	0.87
[14]	LSTM	0.11	0.20	—	—
Our Work (2024)	**Proposed Bi-LSTM**	**0.00198**	**0.04459**	**0.0264**	**0.93735**

4 Conclusion and Future Work

Early crop yield predictions are vital for food security. They enable better agricultural planning and resource allocation. Accurate forecasts help policymakers make data-driven decisions on crop production and pricing, ensuring economic stability. AI-driven insights support precision agriculture, allowing farmers to optimize productivity. This research also helps assess climate change's impact on agriculture, aiding in risk mitigation and sustainability. In this context, the study focuses on crop yield prediction in Bangladesh using deep learning. The goal is to improve agricultural planning by forecasting crop yields. This helps with decision-making for cultivation, imports, and exports. The study considers factors like weather, soil conditions, and crop-water relationships. Four deep learning models were used: Bi-LSTM, LSTM, GRU, and 1D-CNN. Among these, the proposed custom Bi-LSTM achieved the highest precision, followed by LSTM, GRU, and 1D-CNN. Bi-LSTM outperformed the others with the lowest error rates (MSE, RMSE, and MAE) and the highest R^2 score. This indicates its ability to capture temporal dependencies for crop yield prediction. In the future, adding real-time satellite imagery, soil composition, and crop disease data could improve accuracy. Expanding the study to more regions or countries with similar agricultural conditions would enhance adaptability. A real-time crop yield prediction system using live weather and soil sensor data could provide continuous insights. This would help farmers adapt to changing conditions and improve crop management.

References

1. Aboelkhair, H., Morsy, M., El Afandi, G.: Assessment of agroclimatology nasa power reanalysis datasets for temperature types and relative humidity at 2m against ground observations over egypt. Adv. Space Res. **64**(1), 129–142 (2019). https://doi.org/10.1016/j.asr.2019.03.032, https://www.sciencedirect.com/science/article/pii/S0273117719302133
2. Ahmed, F., Bijoy, M., Noori, S.: Smart aquaculture analytics: Enhancing shrimp farming in Bangladesh through real-time iot monitoring and predictive machine learning analysis. Heliyon (2024). https://www.cell.com/heliyon/pdf/S2405-8440(24)13361-0.pdf
3. Ahmed, F., Das, A., Zubair, M.: A machine learning approach for crop yield and disease prediction integrating soil nutrition and weather factors. In: IEEE Conference Paper (2024). https://arxiv.org/pdf/2403.19273
4. Arumuga Arun, R., Saraswathi, T., Meeha, D., Sugeerthi, G.: A machine learning-based agricultural yield forecasting system for predicting crop/plant yield before planting the crops/plants. In: 2023 12th International Conference on Advanced Computing (ICoAC), pp. 1–6 (2023). https://doi.org/10.1109/ICoAC59537.2023.10249364
5. Bangladesh Bureau of Statistics: Bangladesh bureau of statistics (bbs) (2024). https://bbs.gov.bd/
6. Bangladesh Meteorological Department: Bangladesh meteorological department live weather service (2024). https://live6.bmd.gov.bd/. Accessed 05 Oct 2024
7. Bharti, S., Kaur, P., Singh, P., Madhu, C., Garg, N.: Crop yield prediction using cnn-lstm model. In: 2022 IEEE Conference on Interdisciplinary Approaches in Technology and Management for Social Innovation (IATMSI), pp. 1–5 (2022). https://doi.org/10.1109/IATMSI56455.2022.10119308
8. Ifty, R., Hossain, A., Ismail, M.: Enhancing precision agriculture with machine learning & iot: Svr and dtr comparative analysis. In: IEEE International Conference on Smart Farming (2024). https://ieeexplore.ieee.org/abstract/document/10561658/
9. Islam, A., Sultana, T., Hasan, M.: Bangladeshi paddy yield estimation by deep learning approach. In: IEEE International Conference on Computational Intelligence (2024). https://ieeexplore.ieee.org/abstract/document/10725943/
10. Islam, T., Mazumder, T., Roni, M., Nur, M.: A comparative study of machine learning models for predicting aman rice yields in Bangladesh. Heliyon (2024). https://www.cell.com/heliyon/pdf/S2405-8440(24)16795-3.pdf
11. Jhajharia, K., Mathur, P., Jain, S., Nijhawan, S.: Crop yield prediction using machine learning and deep learning techniques. Procedia Comput. Sci. **218**, 406–417 (2023). https://doi.org/10.1016/j.procs.2023.01.023. https://www.sciencedirect.com/science/article/pii/S1877050923000236
12. Meem, R., Turna, T.: Crop yield prediction in Bangladesh: a hybrid machine learning and dnn approach. In: IEEE International Conference on Agricultural AI (2024). https://ieeexplore.ieee.org/abstract/document/10796813/
13. Rahman, F., Khan, M., Tasneem, Z.: Soil classification and crop cultivation prediction: a comparative study of machine learning models. Int. J. Agric. Data Sci. (2024)
14. Saini, P., Nagpal, B.: Deep-lstm model for wheat crop yield prediction in India. In: 2022 Fifth International Conference on Computational Intelligence and Communication Technologies (CCICT), pp. 73–78 (2022). https://doi.org/10.1109/CCiCT56684.2022.00025

15. Shamsuzzoha, M., Shaw, R., Ahamed, T.: Machine learning system to assess rice crop change detection from satellite-derived rgvi due to tropical cyclones using remote sensing dataset. Remote Sens. Appl. (2024), https://www.sciencedirect.com/science/article/pii/S235293852400065X

16. Sudhamathi, T., Perumal, K.: A novel hybrid dnn-rnn framework for precise crop yield prediction. Int. J. Syst. Assur. Eng. Manag. (2024). https://doi.org/10.1007/s13198-024-02577-4

Group Privacy in AI Systems

Ying Li$^{(\boxtimes)}$ and Rohan Gopal Kulkarni

Sunnyvale, USA

lenayingli@gmail.com, rohan.kulkarni1998@gmail.com

Abstract. As artificial intelligence (AI) continues to advance, it brings with it significant privacy and ethical challenges—especially concerning group privacy. Traditional privacy frameworks have primarily focused on protecting individual privacy, but they struggle to address the complexities posed by AI systems that analyze and infer sensitive information from entire groups. This paper explores the unique risks to group privacy in AI, highlighting how AI's ability to draw conclusions about communities, demographics, and social groups can lead to unintended discrimination, stereotyping, and privacy violations. We discuss various technical solutions, such as differential privacy, federated learning, and privacy-preserving technologies (PETs), that aim to safeguard group data while still enabling meaningful analysis. Additionally, we examine ethical considerations around fairness, accountability, and transparency, and propose guidelines to ensure that AI systems respect the privacy of groups. Finally, we discuss future research directions focused on improving AI systems' ability to protect group privacy and developing more robust ethical frameworks for AI deployment across diverse sectors. By integrating both technological advancements and ethical principles, we believe AI can be harnessed to drive positive change without compromising the privacy and dignity of entire communities.

Keywords: AI Ethics · Group Privacy · AI Privacy · Algorithmic Fairness and Bias

1 Introduction

As AI continues to be embedded in many areas of daily life, privacy concerns are becoming more significant. Although AI brings many advantages, it also raises important issues such as data security, consent, potential biases in algorithms, and the need for new laws to meet these challenges in the age of AI [1]. Individual privacy within AI has been studied more frequently, but group privacy has not received the same level of attention [2]. In this paper, we will look at concerns surrounding group privacy, the reasons behind them, and potential solutions.

Y. Li and R. G. Kulkarmi—Independent Researchers.

Y. Li and R. G. Kulkarmi—These authors contributed equally to this work.

Y. Folajimi et al. (Eds.): SIAI 2025, CCIS 2599, pp. 72–86, 2025.
https://doi.org/10.1007/978-3-031-98949-0_8

2 Definition of Group Privacy

Group privacy in AI refers to the protection of sensitive information that pertains to a group as a whole, rather than individual members. This can include a wide range of data, such as location patterns, demographic information, social behaviors, or any other characteristics that define the group. For example, groups might be formed based on shared ethnicity, birthplace, common work habits, or similar interests. The concept of group privacy becomes particularly important when sensitive patterns or trends emerge from collective data, which could expose vulnerabilities or lead to unfair treatment. Safeguarding group privacy ensures that the privacy of the collective identity is respected and that groups are not unintentionally discriminated against or targeted due to the aggregated data about them.

3 Group Privacy vs Individual Privacy

In the broader context of AI privacy, the focus is primarily on protecting personal data and information when using artificial intelligence systems. However, there is an important distinction between individual privacy and group privacy within AI systems. While individual privacy concerns the rights and protection of single users, group privacy takes a different approach, concentrating on the collective interests of groups as a whole [3]. This distinction becomes crucial as AI technologies are increasingly used to analyze and make decisions about larger groups or entire communities. Here's a deeper look into the unique aspects of group privacy:

3.1 Scope of Protection

Group privacy shifts the focus from individuals to the collective rights and interests of groups. This becomes especially important when AI systems are tasked with analyzing and making inferences about entire demographics, communities, or social groups. Rather than just considering individual behavior, group privacy addresses privacy risks that arise when AI identifies patterns, trends, or characteristics of specific groups. For instance, when AI systems track and analyze data from entire neighborhoods, ethnic groups, or social classes, they raise significant concerns about how the privacy of the entire group can be preserved. Group privacy is, therefore, concerned with protecting the group's collective identity, not just the privacy of its individual members [2].

3.2 Inference and Stereotyping

One of the key concerns with group privacy is the ability of AI to infer sensitive group characteristics from seemingly harmless or unrelated data. This issue is especially concerning when such inferences lead to stereotyping or discrimination against specific groups. For example, AI systems might analyze patterns in shopping habits or social media activity and make assumptions about a group based on this data. The danger lies in the potential for these assumptions to reinforce harmful stereotypes or create biased outcomes. Furthermore, algorithmic bias can disproportionately affect certain groups, leading to unequal treatment or outcomes [4]. The privacy risks are not just about data being exposed but also about how that data can be used to create harmful narratives or decisions about an entire group.

3.3 Legal Recognition

Unlike individual privacy, which is well-defined and explicitly protected by existing laws, group privacy is often left unaddressed by current legal frameworks. Most privacy laws are designed to protect the rights of individuals, focusing on the protection of personal information and consent on a case-by-case basis. However, group privacy does not always receive the same level of attention. AI systems often create groups ad hoc by profiling and categorizing data, but these groupings do not necessarily align with recognized legal definitions or rights. This lack of legal recognition means that groups, especially those formed or identified by AI systems, may not have the same legal protections that individuals do. As a result, the rights of these groups are often overlooked or ignored in legal contexts [5].

3.4 Challenges in Consent and Control

Group privacy presents unique challenges when it comes to obtaining meaningful consent and giving groups control over their data. Unlike individual privacy, where a person can explicitly consent to how their data is used, it is often impractical to obtain consent from an entire group. Additionally, members of a group may not even be aware that they are being categorized together by AI systems. This lack of awareness can make it difficult for individuals to understand how their data is being used in the context of group profiling. Even more troubling is that groups, as a collective entity, often have little ability to control how they are profiled or categorized. This raises significant questions about how groups can exercise their privacy rights and ensure they are not unfairly exploited by AI systems [6].

3.5 Ethical Considerations

Group privacy also brings with it complex ethical dilemmas that are not as prominent in individual privacy concerns. One key issue is balancing the privacy rights of individuals within a group against the potential benefits to society that might arise from using group data. For example, AI systems that aggregate group data could be used for research or public good, but they could also compromise the privacy of the group as a whole. Furthermore, there is an ongoing tension between protecting privacy and ensuring that data can be used for beneficial purposes like improving public services, advancing scientific research, or addressing social issues [7]. This raises important questions about the ethical responsibility of AI systems to consider the long-term societal impacts of group profiling, particularly when such profiling could lead to stigmatization or exploitation of certain groups.

Group privacy in AI differs from traditional individual privacy approaches in several important ways. By addressing the specific concerns related to group data and collective rights, group privacy seeks to protect the interests of groups while preventing AI-driven discrimination, stereotyping, and manipulation. While individual privacy remains a crucial aspect of AI ethics, it is equally important to consider how AI systems affect entire communities or demographic groups [8]. Balancing these interests ensures a more holistic approach to privacy in the age of AI, where both individuals and groups are protected from harm.

4 The Mechanisms by Which AI Systems Infer Group Characteristics

AI systems are increasingly capable of inferring group characteristics through various advanced mechanisms. These capabilities, while powerful, can raise significant privacy and ethical concerns. Here are some key methods through which AI systems make these inferences:

4.1 Complex Pattern Detection

Deep learning models are particularly adept at detecting intricate patterns in data that may not be immediately obvious to humans. These sophisticated models can sometimes reveal group-level information, even when that information was not explicitly provided. For example:

- Neural networks are capable of picking up subtle linguistic cues in text that could be correlated with demographic characteristics such as sex or age [9].
- Similarly, computer vision algorithms can detect visual features in images that are indicative of group membership, such as ethnicity or socioeconomic status [10]. These types of features might not be directly labeled in the data, but can still be inferred with surprising accuracy.

4.2 Proxy Variables and Indirect Inference

AI systems often rely on proxy variables' data points that, while seemingly unrelated, correlate with protected attributes. This enables AI to make inferences about individuals or groups without needing explicit demographic information. Examples of such indirect inferences include:

- Shopping habits or location data might serve as proxies for sensitive attributes such as socioeconomic status or race [11].
- Behavioral patterns, such as browsing history or content preferences, can sometimes reveal characteristics such as sexual orientation or political affiliation, even when these traits are not directly disclosed [12].

4.3 Aggregation and Statistical Analysis

When large data sets are aggregated, AI systems can identify trends or patterns specific to certain groups or demographics. Through statistical techniques, such as clustering, AI can group individuals based on shared attributes, sometimes uncovering sensitive characteristics. For example:

- Aggregating individual-level data can help reveal patterns that are specific to particular demographics, even if the data does not explicitly include demographic markers [13].
- Clustering techniques can group users with similar attributes, which may inadvertently expose protected characteristics such as race, ethnicity, or income level [14].

4.4 Latent Variable Modeling

Advanced AI techniques, such as deep latent variable models, can infer hidden or unobserved factors that explain patterns in the data. These models can use noisy or indirect data points to uncover sensitive group attributes. For example:

- Deep latent variable models can use proxy variables to make educated guesses about unobserved confounders, such as income level or health conditions, that might indirectly relate to group characteristics [15].
- These models are capable of capturing complex relationships between observable features (like online behavior or health data) and latent characteristics that are tied to specific groups.

4.5 Interaction Effects and Non-Linear Relationships

Machine learning algorithms, particularly deep learning models, excel at identifying complex interactions between multiple variables. These non-linear relationships can reveal group-level information that might not be apparent from analyzing individual variables alone. For example:

- The interaction between input features, such as shopping patterns and social media activity, may uncover group-level insights that are hidden in isolated data points, such as income or political alignment [11].
- The combination of various data types (e.g., online activity, location, and purchase history) can create unexpected proxies for protected attributes, leading to unintended disclosures of sensitive group characteristics.

Through these mechanisms, AI systems can infer powerful insights about group characteristics, often without explicit data on those characteristics. This capability raises serious privacy concerns, particularly when these inferences are used in decision-making processes that impact individuals or entire communities. The ability to infer sensitive information from seemingly unrelated data necessitates careful consideration of the ethical implications, especially as AI is increasingly used to make decisions that affect real-world outcomes.

5 Case Studies of Group Privacy Violations or Potential Risks in Various Domains

Several notable case studies illustrate the ways in which AI systems have violated group privacy, often in significant and troubling ways. These examples show the potential for AI to infringe on collective rights, from healthcare to employment, social media, and biometric data collection. Here are some key cases:

5.1 Healthcare Data Misuse

A prominent case highlighting group privacy violations in healthcare is Dinerstein v. Google (2019) [16]. In this instance, the University of Chicago and several other healthcare organizations shared a vast amount of patient medical records with DeepMind, a Google-owned AI company, for data mining and machine learning purposes. These records were used to train Google's AI systems for diagnostic and search algorithms, potentially breaching the Health Insurance Portability and Accountability Act (HIPAA) regulations, which safeguard patient data. Although the lawsuit was dismissed, the case raised alarms about how easily healthcare data can be accessed and used without patients' explicit consent. The impact was not limited to a few individuals but affected large groups of patients whose personal health information was exposed to AI systems without adequate protections in place.

5.2 Facial Recognition and Biometric Data

AI's ability to collect and process biometric data without consent has led to significant privacy violations, particularly in the context of facial recognition. For example, IBM faced legal action for using publicly available photos from platforms like Flickr to train its facial recognition software [16]. While the images were accessible online, the company did not obtain consent from the individuals featured in the photos, leading to a lawsuit under Illinois' Biometric Information Privacy Act (BIPA). This case underscores how AI can infringe on group privacy by using publicly available images for purposes like surveillance and data profiling without informing or seeking permission from those whose data is being collected.

Similarly, Clearview AI faced multiple lawsuits after scraping billions of facial images from social media platforms without consent [17]. This data was used to create a massive facial recognition database for surveillance purposes. The company was found to have violated privacy laws in several countries, including Canada, where it had collected images of both adults and children for mass surveillance, further violating the group privacy of people who had no idea their faces were being harvested for such purposes.

5.3 Algorithmic Discrimination in Employment

The use of AI in hiring practices has raised concerns about algorithmic discrimination, which can also lead to group privacy violations. In the case of Aon's Adaptive Employee Personality Test (ADEPT-15), the American Civil Liberties Union (ACLU) argued that the algorithmic test disproportionately affected neurodivergent individuals and those with mental health conditions, such as depression or anxiety [16]. The test's cognitive assessments and video screener were found to potentially discriminate based on race, disability, and mental health, affecting entire groups of job applicants. This example illustrates how AI-driven hiring tools, designed to assess candidates' personalities and abilities, can inadvertently impact certain demographic groups, compromising their privacy and potentially leading to biased employment decisions.

These case studies collectively demonstrate the range of ways AI systems can violate group privacy rights. From the misuse of healthcare data and facial recognition technology to algorithmic bias in employment and political manipulation via social media, these examples highlight a critical need for stronger regulations and ethical standards in AI development and deployment. Protecting group privacy is not just about safeguarding individual data but ensuring that whole communities and social groups are not unfairly targeted, discriminated against, or manipulated by AI systems.

6 The Limitations of Current Privacy Protection Frameworks in Addressing Group Privacy

Current privacy protection frameworks are facing significant challenges when it comes to addressing the privacy risks posed by AI systems, especially in relation to group privacy. While these frameworks were designed to protect individual privacy, they often fall short when it comes to safeguarding the privacy of entire communities or groups. Below are some key areas where existing privacy laws fail to address these issues:

6.1 Inadequacy of the Notice-And-Consent Model

The traditional notice-and-consent model, which many privacy laws are built upon, is becoming less effective in the age of AI, especially when it comes to protecting group privacy. This model works by having individuals consent to how their data is used, but it struggles when applied to large-scale AI systems that analyze collective data. For example, systems like smart traffic signals or the data generated by self-driving cars may affect entire communities or groups, but it's nearly impossible to obtain explicit consent from every member of those groups [18]. The traditional approach is ill-suited to address the privacy risks that arise when decisions are made based on the collective data of entire populations.

6.2 Insufficient Focus on Group Privacy

Most existing privacy laws focus on protecting individual privacy, leaving group privacy largely unaddressed. AI systems, however, often analyze data at a much larger scale and can infer sensitive information about groups, leading to potential harms such as stereotyping and discrimination on a broad scale [3]. Current laws simply aren't designed to handle these collective risks. For instance, AI could be used in hiring decisions or to allocate healthcare resources, but without sufficient protections for group privacy, these systems could inadvertently harm entire communities by unfairly disadvantaging specific demographics [19]. Existing frameworks fail to account for these systemic impacts, which are becoming more significant as AI technology evolves.

6.3 Limited Scope of Data Minimization

Data minimization is a key principle in privacy laws that aims to limit the amount of personal data collected to what is absolutely necessary. However, this principle faces challenges when applied to AI systems that require vast amounts of data to be effective. This is especially true when AI systems analyze group-level data, such as in cases of demographic profiling or predictive analytics. The tension between the need for data to improve AI models and the desire to limit data collection to protect group privacy creates a significant challenge for current laws [19]. Existing privacy frameworks often fail to strike the right balance between these two needs, which can lead to over-collection of sensitive group data.

6.4 Lack of Accountability for Algorithmic Decision-Making

One of the biggest concerns with AI systems is the lack of transparency in their decision-making processes. Many AI systems operate as "black boxes", making it difficult to understand how decisions are being made, especially when these decisions impact groups rather than individuals. Current privacy laws don't do enough to address this lack of transparency, which is critical when decisions affecting groups are made. For example, if an AI system disproportionately impacts a particular community in a hiring process or a healthcare setting, there is often no clear path to holding the system accountable for these biases [20]. The existing legal frameworks don't adequately address the need for accountability in group-level decision-making, leaving groups vulnerable to unfair treatment and discrimination.

6.5 Challenges in Defining and Protecting Sensitive Group Information

AI's ability to infer sensitive group information from seemingly innocuous data raises challenges when it comes to defining what constitutes "sensitive" data in the context of groups. Traditional privacy laws were designed to protect personal information, but AI can now analyze large datasets and infer sensitive characteristics, such as race, socioeconomic status, or even political affiliation, indirectly. This makes it harder to determine what information should be protected, particularly when the data is being used to make decisions that affect entire groups, such as predictive policing or credit scoring [19]. Current frameworks were not built to deal with these emerging risks associated with group-level data, and they struggle to keep up with rapid advancements in AI, such as facial recognition and predictive analytics [17].

6.6 Cross-Border Data Flows and Jurisdictional Issues

AI systems are global in nature, which presents significant challenges for privacy laws that are often limited to one jurisdiction. As AI systems analyze data across borders, the privacy risks to groups become harder to govern. For instance, group data may be transferred across jurisdictions with differing privacy standards, putting entire communities at risk. The inability of current privacy frameworks to regulate these cross-border data flows effectively complicates the protection of group privacy on a global scale [20]. Existing laws may not have the reach to protect the privacy of groups across different countries, leading to gaps in privacy protection.

To address these limitations, there is growing recognition of the need for privacy laws that move beyond individual protections and account for the unique risks to group privacy. We need comprehensive approaches that take into account the societal impacts of AI, incorporate privacy-by-design principles from the outset, and develop AI-specific regulations that evolve alongside the technology. This shift in focus will ensure that privacy protections evolve with the rapid advancements in AI and continue to protect both individuals and groups from harm [18–20].

7   Potential Technical Solutions

Several promising solutions have been proposed to address privacy concerns in AI systems, especially when it comes to protecting group privacy. These approaches aim to safeguard the privacy of entire communities or groups, allowing AI systems to be used effectively without compromising collective interests. Here's a look at some of the key methods:

7.1   Differential Privacy (DP)

Differential Privacy (DP) is a technique designed to protect the privacy of groups by adding statistical noise to data or model outputs. This makes it difficult for anyone to infer sensitive information, whether about individuals or entire groups [21, 22]. The beauty of DP is that it ensures privacy while still enabling useful analysis. For example, in federated learning, DP can be applied so that data shared across multiple devices or nodes doesn't expose sensitive group details, allowing for safe collaboration across distributed systems [22].

7.2   Federated Learning (FL)

Federated Learning (FL) is another method designed to protect group privacy. It allows AI models to be trained across multiple devices without sharing raw data. This means that group data can remain on local devices, such as smartphones or IoT devices, while still being used to build a common model. This is especially useful for protecting the privacy of groups because it ensures that no single device or participant ever has access to the full set of sensitive group data [23]. However, while FL is powerful, it's important to combine it with other privacy measures, as attackers may still target model updates that involve sensitive group information [24].

7.3   Data Anonymization and Pseudonymization

Data anonymization and pseudonymization help protect the identities of individuals within groups, which also contributes to group privacy. Anonymization removes or hides personal identifiers from data, making it impossible to trace back to any individual within the group [21]. Pseudonymization, on the other hand, replaces real identifiers with artificial ones, further protecting individuals without compromising the usefulness of the data [25]. Advanced techniques like k-anonymity, l-diversity, and t-closeness make it even harder to uncover sensitive group characteristics, helping to prevent discriminatory profiling or misuse of group data [21].

7.4 Privacy-Preserving Technologies (PETs)

There are several cutting-edge privacy-preserving technologies (PETs) that are particularly helpful in protecting both individual and group privacy:

- Homomorphic Encryption lets you perform computations on encrypted data without exposing it. This is especially useful for group data because it allows for analysis without risking group privacy [21].
- Secure Multi-Party Computation (SMPC) enables different parties to analyze group data together without any of them seeing the raw data. This is valuable when organizations need to collaborate on group-level data analysis but want to protect the privacy of all parties involved [23].
- Zero-Knowledge Proofs allow one party to prove that a computation is correct without revealing any of the underlying data, making them a great tool for group privacy in collaborative settings [26].

7.5 Group Privacy Protections

Emerging research is exploring ways to specifically protect group privacy. For instance, d-privacy is a form of differential privacy that ensures group data stays private while maintaining its overall structure. This helps avoid unfair treatment of groups and ensures that sensitive data remains protected [27]. Another approach, Group Distributionally Robust Training, combines differential privacy with training methods that help AI systems treat different groups fairly, minimizing any performance bias across groups [28].

7.6 Data Minimization and Retention Policies

To further protect group privacy, AI systems should follow data minimization principles, collecting only the data that's absolutely necessary for the system to work. This helps avoid over-collection of sensitive group data [25]. In addition, strict data retention policies should be in place, ensuring that group data isn't kept longer than necessary, further reducing privacy risks for entire communities or groups [26].

By combining these advanced technical solutions with strong ethical guidelines and data governance policies, organizations can significantly improve group privacy protections in AI systems. This balanced approach ensures that AI technologies can still provide valuable insights, while safeguarding the collective privacy of both individuals and groups.

8 Ethical Considerations and Proposed Guidelines

Protecting group privacy in AI systems comes with its own set of unique ethical challenges, especially given AI's ability to infer and act on collective data [29]. Unlike individual privacy, which focuses on protecting personal rights, group privacy involves safeguarding the rights and interests of entire groups—whether those groups are defined by demographics, communities, or social networks [30]. When AI systems analyze data from these groups, there's a risk of harm, discrimination, or stereotyping. Here are some ethical guidelines for addressing group privacy:

8.1 Group Fairness and Non-Discrimination

AI systems need to be designed in a way that prevents harmful biases and discriminatory practices against entire groups. To protect group privacy, it's essential that AI models are assessed for:

– Fairness Across Groups: Ensuring that AI systems don't disproportionately harm certain communities based on characteristics like race, gender, or socioeconomic status [31].
– Mitigating Stereotyping: Preventing AI from making assumptions or generalizations that could lead to unfair profiling or discrimination against specific groups [32].

8.2 Group-Level Transparency and Accountability

Just like individual privacy, group privacy requires transparency. Ethical guidelines should ensure that AI systems disclose how they handle group data. This includes:

– Clear communication about how group data is collected, processed, and analyzed. It's crucial that AI systems inform stakeholders—especially vulnerable groups—about how their collective data is being used [33].
– Explaining AI-driven decisions at the group level, especially when those decisions affect whole communities, such as in healthcare, urban planning, or criminal justice [34]. Transparency in AI's decision-making process helps maintain accountability.

8.3 Data Aggregation and Privacy Risks

AI systems often aggregate data in ways that expose sensitive group characteristics. Ethical guidelines should prioritize:

– Minimizing the collection of sensitive group-level data that could reveal or exploit vulnerabilities within a group. For example, location data might expose patterns about marginalized communities, or health data could disclose group vulnerabilities [35].
– Data Minimization: Ensuring that only the essential group-level data needed for the AI system to function is collected, so privacy isn't compromised [36].

8.4 Group Consent and Collective Control

Obtaining consent for group data presents unique challenges. To address this:

– Develop Collective Consent Mechanisms: Groups should have a say in how their collective data is used. For instance, communities should be informed and consent to how their data is analyzed, especially when it impacts their rights or social standing [37].
– Empower Groups to Control Their Data: Allow groups to opt out of data uses that may involve sensitive information or could harm the group's privacy [38].

8.5 Privacy by Design for Group Data

Just as individual privacy should be built into the design of AI systems, group privacy should be part of the design process as well. This includes:

– Incorporating privacy-enhancing technologies (PETs) like differential privacy and federated learning, which protect collective group data while still allowing for meaningful analysis [39].
– Conducting Group-Level Privacy Impact Assessments: Identifying potential risks to group privacy early in the AI development process is key to ensuring privacy is respected [40].

8.6 Mitigating Group Profiling and Discrimination

AI can sometimes reinforce social divisions by profiling entire groups based on their characteristics. Ethical guidelines must:

– Prevent the Misuse of Group Data for Profiling or Surveillance: AI models should not be used to create discriminatory profiles based on characteristics like ethnicity, income, or location [41].
– Promote Fairness in Group-Level Decisions: AI should not use group characteristics to unfairly discriminate against certain populations, especially those that are already marginalized [34].

8.7 Long-Term Societal Impacts

AI's decisions based on group data can have lasting effects on society. Ethical guidelines should take into account:

– The societal consequences of group profiling, especially when it leads to stigmatization or exclusion. For instance, predictive tools used in policing or hiring could disproportionately affect minority groups, leading to long-term discrimination [35].
– The Ethical Implications of AI in Societal-Level Decisions: AI should be used in ways that promote social cohesion and don't exacerbate inequalities between groups [36].

8.8 Group-Level Auditing and Governance

To ensure group privacy is respected, AI systems must undergo regular audits for compliance with ethical standards. This includes:

– Establishing independent oversight bodies that evaluate the impact of AI systems on group privacy and ensure adherence to ethical guidelines [37].
– Regularly assessing AI models for unintended biases or privacy violations that could affect groups, ensuring ongoing protection for collective privacy rights [34].

By integrating these ethical guidelines into AI development and deployment, we can ensure that both individual and group privacy rights are respected. This balanced approach to AI ethics helps ensure that AI technologies serve the greater good while protecting the privacy and dignity of communities as a whole.

9 Conclusion and Future Work

To wrap things up, AI has incredible potential to drive innovation and improve decision-making, but it also brings some serious privacy and ethical challenges—especially when it comes to group privacy. Traditional privacy frameworks have mostly focused on protecting individuals, and while they've been effective in many cases, they aren't quite enough to handle the complexities that arise when AI analyzes data about groups. AI's ability to infer sensitive information about entire communities or social groups can sometimes lead to unintended consequences like discrimination or stereotyping, which raises valid concerns.

We've seen that there are some promising solutions—like differential privacy, federated learning, and privacy-preserving technologies (PETs)—that can help protect both individuals and groups. But we also need to ensure that ethical principles, such as fairness, transparency, and accountability, are built into the design of AI systems. It's crucial that we create frameworks that allow groups to have a say in how their data is used, empowering communities with the right to control their collective privacy.

The road ahead will involve not just refining these technical solutions, but also developing comprehensive laws and regulations that address the specific risks AI poses to group privacy. We need more research into how to assess the group-level impact of AI, how to improve AI transparency, and how to hold these systems accountable when they affect entire communities. It's clear that AI will continue to play a major role in our future, and we need to make sure it's done in a way that respects the privacy of everyone—not just individuals, but entire groups and communities.

Looking ahead, the focus should be on making AI systems more inclusive and fairer. It's no longer just about protecting individual data; it's about ensuring that AI doesn't inadvertently compromise the privacy of entire groups. By combining the right technology with strong ethical guidelines, we can create a future where AI drives positive change without jeopardizing the privacy and dignity of vulnerable communities. To make this happen, our future research will focus on improving methods like differential privacy and federated learning for group data, ensuring they're better equipped to handle the complexities of collective information. We also plan to explore AI transparency tools that help people understand how their group data is being used, keeping decision-making accountable. Additionally, we'll look into developing scalable ethical guidelines that can be applied across industries like healthcare and education to better protect group privacy in real-world AI applications. Through these efforts, we aim to ensure that AI not only helps us innovate but also respects and protects the privacy of entire communities.

References

1. Gomstyn, A., Jonker, A.: Exploring privacy issues in the age of AI. IBM (2024). https://www.ibm.com/think/insights/ai-privacy/. Accessed 13 Jan 2025
2. Majeed, A., Khan, S., Hwang, S.O.: Group privacy: an underrated but worth studying research problem in the era of artificial intelligence and big data. Electronics **11**(9), 1449 (2022). https://doi.org/10.3390/electronics11091449
3. Sullivan, M.: Examining Privacy Risks in AI Systems. Transcend (2023). https://transcend.io/blog/ai-and-privacy. Accessed 13 Jan 2025

4. Office of the Victorian Information Commissioner (OVIC): Artificial Intelligence and Privacy: Issues and Challenges. OVIC (2018). https://ovic.vic.gov.au/privacy/resources-for-organisat ions/artificial-intelligence-and-privacy-issues-and-challenges/. Accessed 17 Jan 2025
5. Voorzitterasset: Individual vs Group Privacy. IThappens.Nu (2016). https://www.ithappens. nu/individual-vs-group-privacy/. Accessed 13 Jan 2025
6. Rijmenam, M.V.: Privacy in the Age of AI: Risks, Challenges and Solutions. The Digital Speaker (2023). https://www.thedigitalspeaker.com/privacy-age-ai-risks-challenges-soluti ons/. Accessed 13 Jan 2025
7. Guill´en, M.F.: Data by Default: How AI Radically Changes the Data Privacy Landscape. Vision by Protiviti (2024). https://vision.protiviti.com/insight/data-default-how-ai-radically-changes-data-privacy-landscape/. Accessed 13 Jan 2025
8. Landreneau, D.J.: Navigating the Intersection of AI and Data Privacy. Tealium (2023). https://tealium.com/blog/data-governance-privacy/navigating-the-intersection-of-ai-and-data-privacy/. Accessed 13 Jan 2025
9. Uddin, S., Lu, H., Rahman, A., Gao, J.: A novel approach for assessing fairness in deployed machine learning algorithms. Sci. Rep. **14**(1), 17753 (2024). https://doi.org/10.1038/s41598-024-17753-9
10. Kanade, V.: What is Pattern Recognition? Working, Types, and Applications. Spiceworks (2023). https://www.spiceworks.com/tech/artificial-intelligence/articles/what-is-pattern-rec ognition/. Accessed 13 Jan 2025
11. Engelmann, S., Ullstein, C., Papakyriakopoulos, O., Grossklags, J.: What people think AI should infer from faces. In: Proceedings of the 2022 ACM Conference on Fairness, Accountability, and Transparency, pp. 128–141 (2022)
12. Moody, G.: Advanced AI Chatbots Can Now Infer Detailed Personal Attributes from General Social Media Posts. Private Internet Access (2023). https://tinyurl.com/mrw5946b. Accessed 13 Jan 2025
13. Waniek, M., Suri, N., Zameek, A., AlShebli, B., Rahwan, T.: Human intuition as a defense against attribute inference. Sci. Rep. **13**(1), 16088 (2023). https://doi.org/10.1038/s41598-023-16088-y
14. Ferrara, E.: Fairness and bias in artificial intelligence: a brief survey of sources, impacts, and mitigation strategies. Sci **6**(1), 3 (2023). https://doi.org/10.3390/sci6010003
15. Deng, Z., Zheng, X., Tian, H., Zeng, D.D.: Deep causal learning: representation, discovery and inference. arXiv preprint arXiv:2211.03374 (2022)
16. Osman, H.: 7 AI Privacy Violations (+What Can Your Business Learn). Enzuzo (2024). https://www.enzuzo.com/blog/ai-privacy-violations/. Accessed 13 Jan 2025
17. Pearce, G.: Beware the Privacy Violations in Artificial Intelligence Applications. ISACA (2021). https://www.isaca.org/resources/news-and-trends/isaca-now-blog/2021/beware-the-privacy-violations-in-artificial-intelligence-applications/. Accessed 13 Jan 2025
18. Kerry, C.F., et al.: Protecting privacy in an AI-driven world. Brookings Institution (2020)
19. King, J., Meinhardt, C.: Rethinking Privacy in the AI Era: Policy Provocations for a Data-Centric World. Stanford Human-Centered Artificial Intelligence (2024). https://hai.stanford. edu/sites/default/files/2024-02/White-Paper-Rethinking-Privacy-AI-Era.pdf. Accessed 13 Jan 2025
20. Chin-Rothmann, C.: Protecting Data Privacy as a Baseline for Responsible AI. Center for Strategic and International Studies (2024). https://www.csis.org/analysis/protecting-data-pri vacy-baseline-responsible-ai/ Accessed:2025–01–13
21. Rathnayake, D.: Ensuring Privacy in the Age of AI: Exploring Solutions for Data Security and Anonymity in AI. Fortra (2024). https://www.tripwire.com/state-of-security/ensuring-pri vacy-age-ai-exploring-solutions-data-security-and-anonymity-ai/. Accessed 13 Jan 2025

22. Cook, D.: Combining Privacy Preserving Telemetry with Differential Privacy. Divvi Up (2024). https://divviup.org/blog/combining-privacy-preserving-telemetry-with-different ial-privacy/. Accessed 13 Jan 2025
23. Kandati, D.R., Anusha, S.: Security and privacy in federated learning: a survey. Trends Comput. Sci. Inf. Technol. **8**(2), 029–037 (2023)
24. Near, J., Darais, D., Buckley, D., Durkee, M.: Privacy Attacks in Federated Learning. NIST (2024). https://www.nist.gov/blogs/cybersecurity-insights/privacy-attacks-federated-learning. Accessed 13 Jan 2025
25. ITMAGINATION: AI Solutions and Privacy: Overcoming Common Challenges & Committing to Responsible AI. ITMAGINATION (2025). https://tinyurl.com/83drttnr. Accessed 13 Jan 2025
26. DigitalOcean: AI and Privacy: Safeguarding Data in the Age of Artificial Intelligence. DigitalOcean (2025). https://www.digitalocean.com/resources/articles/ai-and-privacy. Accessed 13 Jan 2025
27. Galli, F., Jung, K., Biswas, S., Palamidessi, C., Cucinotta, T.: Advancing personalized federated learning: group privacy, fairness, and beyond. SN Comput. Sci. **4**(6), 831 (2023)
28. Hansen, V.P.B., Neerkaje, A.T., Sawhney, R., Flek, L., Søgaard, A.: The impact of differential privacy on group disparity mitigation. arXiv preprint arXiv:2203.02745 (2022)
29. Barocas, S., Hardt, M., Narayanan, A.: Fairness and Machine Learning. fairmlbook.org (2019)
30. Dastin, J.: Insight - Amazon scraps secret AI recruiting tool that showed bias against women. Reuters (2018). https://tinyurl.com/2fxxnf3h/. Accessed 13 Jan 2025
31. Ribeiro, M.T., Singh, S., Guestrin, C.: "Why should I trust you?" explaining the predictions of any classifier. In: Proceedings of the 22nd ACM SIGKDD International Conference on Knowledge Discovery and Data Mining, pp. 1135–1144 (2016)
32. Mittelstadt, B.D., Allo, P., Taddeo, M., Wachter, S., Floridi, L.: The ethics of algorithms: mapping the debate. Big Data Soc. **3**(2), 2053951716679679 (2016)
33. Shokri, R., Stronati, M., Song, C., Shmatikov, V.: Membership inference attacks against machine learning models. In: 2017 IEEE Symposium on Security and Privacy (SP), pp. 3–18 (2017)
34. Solove, D.J.: Understanding Privacy. Harvard University Press, Cambridge (2010). ISBN: 9780674035072
35. Cavoukian, A.: Privacy by design: The seven foundational principles. IAPP Resource Center (2021)
36. Gentry, C.: Fully homomorphic encryption: what is it and how will it change the future? Commun. ACM **53**(3), 97–105 (2009). https://doi.org/10.1145/1666420.1666444
37. Angwin, J., Larson, J., Mattu, S., Kirchner, L.: Machine Bias. ProPublica (2016). https://www.propublica.org/article/machine-bias-risk-assessments-in-criminal-sentencing. Accessed 1 Jan 2025
38. Eubanks, V.: Automating Inequality: How High-Tech Tools Profile, Police, and Punish the Poor. St. Martin's Press, New York (2018)
39. O'Neil, C.: Weapons of Math Destruction: How Big Data Increases Inequality and Threatens Democracy. Crown (2017)
40. Noble, S.U.: Algorithms of Oppression: How Search Engines Reinforce Racism. New York University Press (2018)
41. Binns, R.: Fairness in machine learning: lessons from political philosophy. In: Conference on Fairness, Accountability and Transparency, pp. 149–159 (2018)

Shaping the Future of Learning: AI-Driven Adaptive Feedback for Programming Education in Resource-Constrained Setting

Yetunde Folajimi[1(✉)], Leonidas Deligiannidis[1], Salem Othman[1], and Shawren Singh[2]

[1] School of Computing and Data Science, Wentworth Institute of Technology, 550 Huntington Avenue, Boston, MA 02115, USA
`{folajimiy,deligiannidisl,othmans1}@wit.edu`
[2] School of Information Science, University of South Africa, Preller Street, Muckleneuk Ridge, Pretoria 0002, South Africa
`singhs@unisa.ac.za`

Abstract. The persistent digital divide continues to hinder equitable access to high-quality education, particularly in low-resource regions where internet connectivity, educator capacity, and personalized learning infrastructure are limited. This paper presents an AI-driven assessment system that advances programming education through adaptive feedback tailored to individual learners, even in offline environments. By integrating GPT for generating contextually relevant programming questions, BERT for automatic difficulty classification, and a reinforcement learning engine for dynamic adaptation, the system offers real-time personalization without requiring continuous internet access. In a controlled evaluation, the system generated a curated set of 500 programming questions, with 92% passing automated filtering, and expert raters awarding an average score of 4.4/5 across technical accuracy, pedagogical relevance, and linguistic clarity. The adaptive engine demonstrated strong responsiveness, adjusting question difficulty within 5 to 10 iterations for simulated learners at beginner, intermediate, and advanced levels. Additionally, the system achieved a low average processing latency of 550 ms, ensuring seamless interaction and learner engagement in real time. What sets this work apart is its commitment to educational equity. Designed specifically for resource-constrained settings, the system's offline functionality allows schools and learning centers with limited infrastructure to benefit from the same level of AI-enhanced learning found in high-resource contexts. By reducing educator workload, scaling individualized instruction, and increasing access to quality programming education, this work exemplifies how AI can be harnessed for social good. It contributes a replicable, scalable model for bridging educational gaps and shaping the future of learning in underserved communities worldwide.

Keywords: AI in Education · Adaptive Learning · Programming Education · Educational Equity · Offline Learning Systems · Reinforcement Learning

---

L. Deligiannidis, S. Othman and S. Singh—These authors contributed equally to this work.

1   Introduction

1.1   Background

Personalized assessments are critical in programming education, where students often struggle with complex and abstract concepts requiring iterative practice and reinforcement. Traditional approaches, such as static question sets, fail to accommodate the diverse learning trajectories of students, leading to disengagement and suboptimal learning outcomes [1]. Adaptive and formative assessments, which adjust to individual proficiency levels and provide timely feedback, have been shown to significantly enhance student engagement and knowledge retention [2].

Advances in Artificial Intelligence (AI), particularly in Natural Language Processing (NLP), have created new opportunities for dynamic and scalable personalized assessments. Large Language Models (LLMs) such as GPT and BERT have demonstrated exceptional capabilities in generating coherent and contextually relevant text, making them ideal candidates for educational content creation. Recent research highlights their potential in automating quiz generation and adaptive feedback, offering a transformative approach to teaching and assessment in computer science education [3, 4]. Despite their promise, integrating such systems into practical educational settings remains a challenge due to issues like limited adaptability and dependence on constant internet connectivity.

1.2   Motivation

A key limitation of many existing AI-driven educational tools is their reliance on constant internet connectivity, making them inaccessible in regions with unreliable or no internet access. By incorporating offline functionality, our system addresses this critical gap, ensuring equitable access to high-quality adaptive assessments. In resource-constrained environments, pre-generated question banks and localized computation using BERT and RL enable seamless operation without sacrificing adaptability or personalization. This capability empowers educators and learners in underserved regions, bridging the digital divide in computer science education.

1.3   Educational Equity and Social Impact

Despite rapid advances in AI-driven education technologies, access to such innovations remains uneven across global learning environments. In many underserved or resource-limited settings, students face significant challenges such as limited access to trained educators, unreliable internet connectivity, and a lack of adaptive learning tools tailored to their needs. These disparities not only affect individual learning out-comes but also exacerbate systemic educational inequalities. This work positions AI as a transformative tool for bridging such gaps by delivering high-quality, adaptive, and personalized programming instruction to learners regardless of location or infrastructure. By prioritizing offline functionality, scalability, and pedagogical rigor, our system is designed to empower marginalized learners and extend the benefits of AI-enhanced education to communities historically excluded from technological advances. This contributes to broader efforts to democratize STEM education and aligns with global goals for inclusive and equitable quality education (e.g., UN SDG 4).

1.4   Contributions

This paper introduces a novel AI-driven system that addresses these challenges by integrating GPT, BERT, and reinforcement learning to create a dynamic and scalable personalized assessment platform. The key contributions of this work include:

- **Dynamic Question Generation:** Leveraging GPT for generating diverse, contextually relevant programming questions tailored to individual learning needs.
- **Difficulty Classification:** Using BERT to categorize questions into easy, medium, and hard levels based on linguistic complexity and cognitive demands.
- **Reinforcement Learning for Adaptation:** Implementing RL to dynamically adjust question difficulty in real-time based on students' performance, ensuring sustained engagement and optimal learning conditions.
- **Offline Deployment:** Developing a system capable of functioning without internet connectivity, expanding access to personalized learning in resource-limited settings.

A key limitation of many existing AI-driven educational tools is their reliance on constant internet connectivity, making them inaccessible in regions with unreliable or no internet access. By incorporating offline functionality, our system addresses this critical gap, ensuring equitable access to high-quality adaptive assessments. In resource-constrained environments, pre-generated question banks and localized computation using BERT and RL enable seamless operation without sacrificing adaptability or personalization. This capability empowers educators and learners in underserved regions, bridging the digital divide in computer science education.

2   Related Work

2.1   Adaptive Learning Systems

Adaptive learning systems have long been a cornerstone of personalized education, tailoring content delivery to individual learners' needs. Early systems such as Cognitive Tutor and Bayesian Knowledge Tracing (BKT) have demonstrated the ability to model student knowledge and adapt content delivery to evolving proficiency levels [5, 6]. These systems leverage cognitive task analysis to respond to diverse learning trajectories effectively. While these systems are highly effective, they rely on predefined question sets and rule-based methods for adaptation. Doroudi et al. [7] highlighted that the evidence supporting the superiority of reinforcement learning (RL) for instructional sequencing over traditional methods is mixed. However, they also note the potential for RL to adapt dynamically in ways that traditional systems cannot, particularly when integrated with nuanced student interaction data. Our work aims to build on this foundation by addressing limitations in content generation and dynamic adaptation.

For example, traditional approaches often depend on static content that cannot fully account for nuanced behavioral patterns or quickly evolving learning needs. In contrast, systems such as ours employ Large Language Models (LLMs) like GPT for dynamic content generation, ensuring a broader range of adaptable, contextually relevant programming questions. Furthermore, while knowledge tracing frameworks have proven effective in predicting student understanding over time, they lack the capacity to incorporate real-time feedback loops driven by reinforcement learning, as demonstrated in our system.

Our system complements these established methods by integrating:

1. **Dynamic Question Generation:** Unlike rule-based systems, which are limited to predefined content, our use of GPT allows for the generation of diverse and contextually relevant programming questions.
2. **Real-Time Adaptation with Reinforcement Learning:** The RL engine dynamically adjusts question difficulty based on real-time performance metrics, providing a more granular and immediate adaptation mechanism than traditional approaches.
3. **Offline Functionality:** While many adaptive systems depend on internet connectivity, our system is designed to function seamlessly offline, addressing the accessibility challenges highlighted in underserved regions.

We recognize that intelligent tutoring systems like those described by Koedinger et al. [6] excel at leveraging cognitive task analysis to infer strategies and adapt to students' evolving proficiency levels. However, our system extends these capabilities by employing modern AI techniques for real-time question adaptation and contextual generation, which are particularly impactful in domains like programming education where the variety and complexity of questions are critical for effective learning.

Table 1 summarizes the key differences between traditional systems and our approach.

By incorporating GPT and BERT for question generation and difficulty classification alongside RL for dynamic adaptation, our system bridges gaps in adaptability and scalability, particularly in environments with limited resources. It builds on the strong foundation laid by earlier systems while addressing their inherent constraints.

Table 1. Comparison of Traditional Systems and Proposed System

Feature	Traditional Systems	Proposed System
Content Generation	Predefined, rule-based	GPT-driven, dynamic
Difficulty Adaptation	Static, rule-based	RL-driven, real-time
Offline Functionality	Limited	Fully offline capable
Domain-Specific Flexibility	Moderate	High, using fine-tuned LLMs
Engagement with Real-Time Feedback	Limited	Strong, using RL and LLM integration

2.2   AI in Education

Large Language Models (LLMs), such as GPT and BERT, have demonstrated significant capabilities in generating text-based content, including educational quizzes. Folajimi [3] benchmarked GPT-3, GPT-4, and BERT for quiz generation, highlighting the strengths of GPT models in generative tasks and BERT in contextual understanding. These models enable the automated creation of diverse and adaptive question sets, reducing educator workload and enhancing the scalability of personalized assessments.

Despite their potential, LLM-based systems face challenges, including the accuracy of generated questions and their alignment with pedagogical goals [8]. Reinforcement learning, when integrated with LLMs, can address some of these gaps by dynamically adjusting content based on real-time performance metrics. Research has also shown that AI-driven intelligent tutoring systems, such as CodeWorkout and AutoGrader [9, 10], enhance learning outcomes through automated assessments and immediate feedback. However, many of these systems depend on static question banks and predefined difficulty levels, limiting their adaptability to diverse student needs.

Recent advancements in programming education have explored the use of AI-driven systems to enhance assessment and learning outcomes. Hou et al. [11] introduced *Codetailor*, a system leveraging Large Language Models (LLMs) to generate personalized Parsons puzzles. This approach emphasizes engagement through tailored problem types, showing promise in supporting learners in real-time by adapting puzzle complexity based on their performance. Similarly, Ma et al. [12] proposed the use of LLMs as teachable agents for debugging, focusing on improving students' debugging skills through interactive and adaptive support mechanisms. Both systems demonstrate the power of integrating LLMs into programming education to personalize and enhance the learning experience.

While these contributions advance the state of the art, they are predominantly designed for online and well-connected environments, relying heavily on real-time access to LLM APIs or cloud infrastructure. Additionally, their focus is primarily on individual aspects of programming education, such as puzzle-solving [11] or debugging [12], without addressing the broader need for holistic assessment systems.

Our work extends these approaches by:

- **Offline Functionality**: Unlike Hou et al. [11] and Ma et al. [12], our system operates seamlessly in offline environments. This capability is critical for resource-constrained settings where internet connectivity may be unreliable or unavailable, ensuring equitable access to adaptive assessments.
- **Integrated AI Techniques**: Our system combines GPT for question generation, BERT for difficulty classification, and reinforcement learning for real-time adaptability. This integration allows for a dynamic adjustment of question difficulty, supporting sustained engagement across diverse learner profiles.
- **Comprehensive Assessment Scope**: Beyond debugging or specific problem types, our system generates diverse question formats (e.g., multiple-choice, fill-in-the-blank, and coding exercises) across programming topics, making it a more comprehensive tool for programming education.

By addressing these gaps, our work complements and builds upon the advancements presented by Hou et al. and Ma et al., offering a scalable, robust, and accessible solution for adaptive programming education.

2.3 Dynamic Assessments

Dynamic assessments are critical for maintaining engagement and optimizing learning outcomes. Traditional systems often lack the capability to adapt to individual student performance in real-time, relying instead on static content generation. This static approach fails to ensure that students are consistently challenged at an appropriate level, potentially leading to disengagement or frustration [1].

AI-driven dynamic assessment systems address these limitations by leveraging reinforcement learning to adjust question difficulty based on real-time feedback. For instance, Matsuda et al. [13] demonstrated the use of RL in math tutoring systems to optimize content delivery, resulting in improved learning outcomes. However, these systems often focus on specific domains and do not generalize well to programming education. Furthermore, existing AI-based tools lack robust evaluation mechanisms to ensure the validity of generated questions. Validity, including alignment with course objectives and technical accuracy, remains an underexplored area in dynamic assessment systems.

By integrating LLMs with reinforcement learning, the system described in this paper bridges these gaps, offering both dynamic adaptability and pedagogical validity. This approach ensures that assessments are not only personalized and engaging but also aligned with educational goals, addressing critical gaps in current adaptive learning technologies.

3 System Design

The system integrates GPT for question generation, BERT for difficulty classification, and reinforcement learning (RL) for dynamic adjustment of question difficulty. Additionally, it supports offline functionality, enabling broader accessibility in resource-constrained environments. Fine-tuning was not pursued in this iteration of the system due to the following considerations:

- **Scalability:** Deploying pre-trained models without fine-tuning reduces computational overhead, making the system viable for offline and resource-constrained environments.
- **Flexibility:** By employing carefully designed prompts for GPT and rule-based mappings for BERT, the system achieves alignment with pedagogical objectives without requiring extensive domain-specific datasets.
- **Accessibility:** Avoiding fine-tuning ensures compatibility with lower-end devices, enabling deployment in underserved regions.

To ensure the quality and relevance of the generated programming questions, we employ a two-stage evaluation pipeline. This process balances automation and expert review to systematically filter and validate the questions before deployment.

- **Automated Filtering Stage:** In the first stage, questions undergo automated checks for:

  - *Syntax Errors:* Ensuring the questions are free from syntactic issues that could affect comprehension or accuracy.

- *Logical Inconsistencies:* Detecting any contradictions or errors in logic that may confuse learners.
- *Content Relevance:* Filtering out questions that do not align with the intended learning objectives or programming topics.

- **Expert Review Stage:** Questions that pass the automated filtering stage are reviewed by subject matter experts to ensure:

  - *Linguistic Clarity:* The language is clear, precise, and accessible for students of diverse backgrounds.
  - *Pedagogical Relevance:* The question aligns with instructional goals and targets the appropriate cognitive level.
  - *Technical Accuracy:* The programming concepts, syntax, and solutions are correct and consistent.

This two-stage pipeline ensures that only high-quality, pedagogically sound questions are delivered to learners. The questions undergo automated filtering for syntax, logic, and relevance, followed by expert review to ensure clarity, relevance, and technical accuracy. The process is summarized in Fig. 1.

3.1 Architecture

The system consists of four main components:

1. **GPT-Driven Question Generation:** Generates diverse and contextually relevant programming questions tailored to educational needs.
2. **BERT-Based Difficulty Classification:** Classifies questions into easy, medium, or hard categories based on complexity and cognitive demands.
3. **Reinforcement Learning Engine:** Dynamically adjusts question difficulty based on real-time student performance.
4. **Offline Functionality Module:** Ensures seamless operation without internet connectivity by pre-generating and storing question sets locally.

A high-level architecture diagram (see Fig. 3) illustrates the data flow between these components. The process begins with GPT generating questions, followed by BERT assigning initial difficulty levels. The RL engine dynamically adjusts subsequent questions based on student performance metrics.

3.2 Offline Deployment for Accessibility

The offline functionality of our system enables equitable access to adaptive learning technologies, particularly in underserved regions with limited infrastructure. By pre-generating a diverse set of questions with GPT and utilizing lightweight local hardware for BERT and RL, the system dynamically adjusts question difficulty without requiring constant connectivity. This ensures seamless operation even in environments with unreliable or no internet access. Hypothetically, in a rural school with intermittent electricity and no internet access, educators could deploy the system on a standalone laptop to

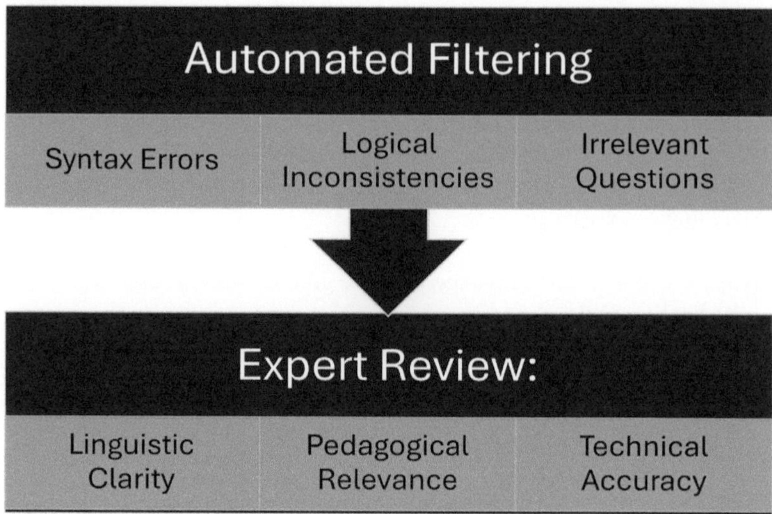

Fig. 1. Two-stage Evaluation Process.

deliver personalized assessments. Upon regaining connectivity, the system can synchronize performance data with central servers for further analysis and refinement, but this is not essential for day-to-day operation.

3.2.1   Question Generation (GPT)

GPT is utilized in its pre-trained state to dynamically generate programming questions. To ensure relevance and alignment with educational goals, carefully structured prompts guide the model to produce diverse question types. For instance:

- **Example Prompt for Java**

   "Generate a multiple-choice question testing the concept of inheritance and polymorphism in Java, with four answer options, including one correct answer."

- **Example Prompt for Python**

   "Create a fill-in-the-blank question about the use of loops in Python programming, including both question text and solution."

3.2.2   Difficulty Classification (BERT)

BERT is employed to classify the difficulty of the generated questions into three categories: *easy*, *medium*, and *hard*. Instead of fine-tuning, the model's contextual understanding is leveraged to evaluate linguistic complexity and cognitive demands. The classification process involves:

1. **Input Processing:** Each generated question is tokenized and input into the pretrained BERT model.

2. **Feature Extraction:** BERT assesses semantic and syntactic elements of the question text, such as vocabulary complexity and required logical reasoning.
3. **Rule-Based Mapping:** Output embeddings are combined with pre-defined thresholds (e.g., sentence length, domain-specific keywords) to assign difficulty levels:

   - **Easy:** Basic syntax and conceptual recall.
   - **Medium:** Logical reasoning or application of intermediate concepts.
   - **Hard:** Higher-order thinking, such as recursion or advanced algorithms.

The BERT model is fine-tuned on a labeled dataset of programming questions with human-annotated difficulty levels, ensuring alignment with expert judgment. Initial classifications are periodically validated using student performance data to refine difficulty assignments. For example, questions frequently answered correctly by students may be reclassified to lower difficulty levels, while more challenging questions may be upgraded.

3.2.3   Reinforcement Learning Engine

The RL framework employs a reward mechanism that optimizes engagement and learning outcomes. It ensures that question difficulty aligns with the learner's Zone of Proximal Development (ZPD), maintaining a balance between challenge and accessibility. Figure 2 provides an overview of the RL reward mechanism.

The RL framework comprises the following components:

- **States:** Represent the student's current proficiency level, determined by metrics such as:

  - *Accuracy:* Percentage of correct responses over recent questions.
  - *Response Time:* The time taken to answer questions.
  - *Engagement:* Metrics such as skipped questions or repeated attempts.

- **Actions:** Correspond to selecting the next question difficulty level (*easy, medium, hard*).

- **Rewards:** Guide the RL agent's decisions to optimize engagement and learning, as detailed below.

Reward Definition: The reward function evaluates the system's choice of question difficulty, balancing engagement and learning. Rewards are defined as follows:

- **Positive Rewards:** Awarded for:

  - *Difficulty Match:* When the selected difficulty aligns with the student's proficiency.
  - *Correct Responses with Effort:* When a correct answer demonstrates sufficient engagement (e.g., appropriate response time).

- **Negative Rewards:** Assigned for:

  - *Overly Easy Questions:* When a question is answered too quickly and effortlessly.

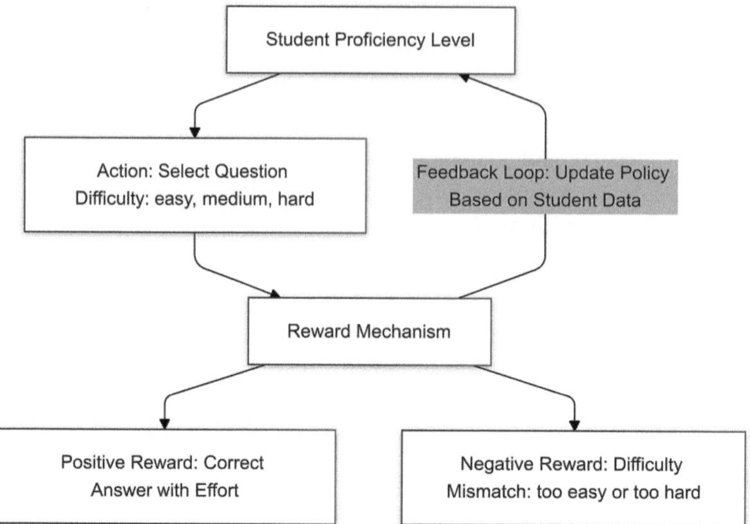

Fig. 2. Reinforcement Learning Reward Mechanism. The system dynamically adjusts question difficulty based on student performance, with feedback loops to refine the RL policy.

- *Excessively Difficult Questions:* When students struggle repeatedly, indicating a mismatch in difficulty.

- **Neutral Rewards:** Applied to inconclusive outcomes (e.g., borderline response times or mixed correctness), preventing abrupt adjustments.

  Reward Function:
  The reward function $R(s, a)$ is defined as:

$$R(s, a) = W_c \cdot C + W_t \cdot T - W_e \cdot E$$

where:

- W_c, W_t, W_e: Weights assigned to correctness (C), response time (T), and engagement metrics (E).
- C: Binary value (1 for correct, 0 for incorrect).
- T: Inversely proportional to response time.
- E: Penalty for skipped or repeated incorrect attempts.

  The RL mechanism follows these steps:

1. **Initial State:** A baseline question difficulty is assigned (e.g., *easy* for beginners).
2. **Performance Monitoring:** The system tracks student metrics (e.g., correctness, response time) to compute rewards.
3. **Dynamic Adjustment:**
   - High-performing students receive progressively harder questions.
   - Struggling students are presented with easier questions or hints.

4. **Feedback Loop:** The reward is fed back to update the RL policy, refining future difficulty selection.

Example Scenario

- **State:** Intermediate proficiency.
- **Action:** Assign a *hard* question.
- **Outcome:** Incorrect response with visible disengagement.
- **Reward:** -10 (difficulty mismatch).

3.3 Offline Functionality

Unlike many AI-driven educational tools that require constant internet connectivity, this system is designed for offline deployment. Key features include:

- **Pre-Generated Questions:** GPT generates a batch of questions in advance, stored locally.
- **Local Execution:** BERT and RL operate entirely on local hardware, processing student interactions and adjusting question difficulty in real-time.
- **Data Synchronization:** When connectivity is available, the system can sync performance data for analytics and updates, but this is not required for core functionality.

This offline capability extends personalized learning to regions with limited infrastructure, ensuring that students can benefit from adaptive assessments regardless of connectivity.

4 Methodology

This study builds on earlier work that benchmarked LLMs for generating programming quiz questions [3]. While the prior study focused on generating large volumes of questions (800 per model) to assess model performance, this work refines the approach by prioritizing the quality, adaptability, and scalability of generated content. To achieve this, we integrate a reinforcement learning (RL) component for dynamic difficulty adjustment and implement a streamlined evaluation pipeline for rigorous validation. To ensure scalability and relevance in simulated testing, we curated a subset of questions that maintain diversity across topics and difficulty levels. This approach streamlines evaluation and avoids redundancy while enabling robust testing of RL adaptability. Compared to our earlier focus on generating large volumes of questions (800 per model), this study optimizes resource use by focusing on a smaller, high-quality subset. The refined pipeline incorporates automated filtering and expert validation, prioritizing precision and rigor over quantity to ensure the generated questions meet both technical and pedagogical standards.

4.1 Question Generation Pipeline

The question generation pipeline begins with tailored prompts guiding GPT to produce programming-related questions. Input prompts include details such as:

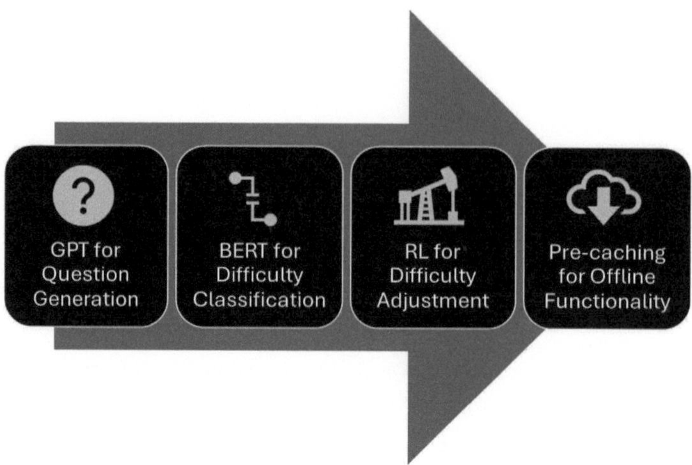

Fig. 3. High-level architecture of the dynamic question generation system.

- Programming topic (e.g., inheritance in Java).
- Desired question format (e.g., multiple choice or short answer).
- Complexity level.

Once generated, the questions are classified by BERT to assign an initial difficulty level. These questions are pre-cached for offline usage and dynamically adapted during student interaction. This ensures accessibility and scalability, even in environments with limited connectivity.

4.2 Reinforcement Learning Simulation

The introduction of reinforcement learning necessitates evaluating how question difficulty adapts dynamically. Rather than increasing the number of questions, we focus on a smaller set of high-quality questions, allowing the RL component to optimize and refine its performance through repeated interactions. Figure 4 outlines the RL process, which consists of:

1. **State Representation**: Student proficiency is represented as the current state, determined by metrics such as correctness of responses and response times.
2. **Actions**: The RL agent selects actions by adjusting the question difficulty to *easy*, *medium*, or *hard*.
3. **Reward Mechanism**: The system receives rewards based on performance alignment. Correct responses to appropriately challenging questions yield positive rewards, while incorrect responses trigger difficulty adjustments.
4. **Feedback Loop**: Student performance data is fed back into the RL agent to update its policy, optimizing future difficulty adjustments.

The RL component is tested using simulated student profiles representing diverse learning behaviors (e.g., beginners, intermediates, and advanced learners). Simulations allow exploration of edge cases, scalability testing, and the validation of RL behavior without relying on real-world participants.

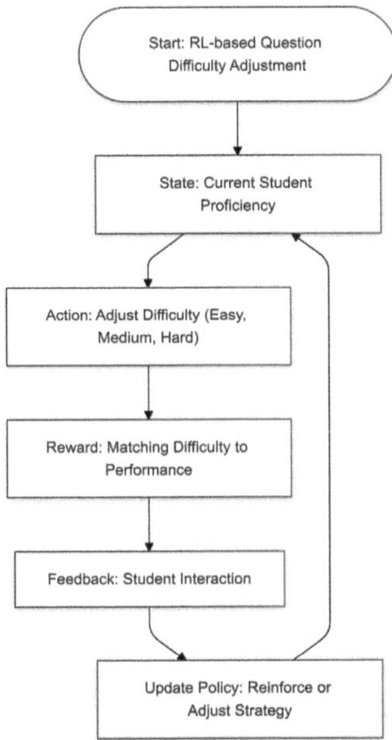

Fig. 4. Reinforcement Learning Flowchart: Dynamic adjustment of question difficulty based on student interaction.

4.3 Evaluation Pipeline

The two-stage evaluation pipeline, shown in Fig. 5, ensures the generated questions are accurate, relevant, and pedagogically sound. The stages are:

1. **Automated Filtering Stage**: Questions undergo initial automated validation to identify syntax errors, logical inconsistencies, or irrelevant content.
2. **Expert Review Stage**: Questions that pass the automated stage are reviewed by programming educators. They evaluate:

   - **Technical Accuracy**: Alignment with programming concepts.
   - **Pedagogical Relevance**: Suitability for educational objectives.
   - **Linguistic Clarity**: Precision and clarity of phrasing.

   The expert panel rates each question on a 1–5 scale. Questions scoring an average of 4.0 or higher are accepted. Feedback from the expert review is iteratively fed back into the question generation pipeline to enhance quality.

4.4 Tools and Technologies

The system is implemented using Python for scripting, data preprocessing, and evaluation. Key tools include:

- GPT and BERT models accessed through OpenAI and Hugging Face APIs.
- Python libraries such as Pandas and SciPy for statistical analysis.

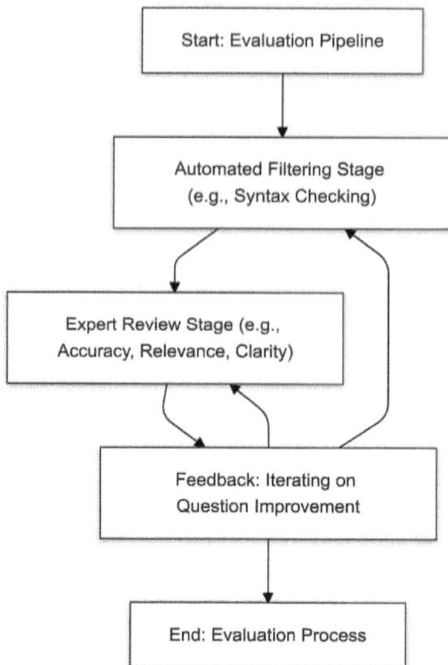

Fig. 5. Evaluation Pipeline: Two-stage validation process with feedback loops for continuous improvement.

- Local deployment for offline functionality, ensuring accessibility in resource-limited settings.

4.5 Evaluation Metrics

The system's performance is evaluated using the following metrics:

1. **Question Quality**: Percentage of valid questions based on expert ratings.
2. **Adaptability**: RL performance, including convergence speed and difficulty stability.
3. **Efficiency**: Time taken to generate, classify, and adjust questions.
4. **Diversity**: Coverage of topics, question types, and difficulty levels.

5 Results

Our results focus are on the quality of generated questions, the adaptability of the reinforcement learning (RL) component, and the system's overall efficiency. The results reflect a refined evaluation process that prioritizes question quality, diversity, and dynamic adaptability.

5.1 Question Quality

The system's question generation and filtering process resulted in a curated subset of 500 high-quality questions, selected after automated filtering and expert validation. This refined focus builds on our prior work, which generated 800 questions per model, but shifts emphasis to a smaller, cleaner dataset for rigorous testing.

- **Automated Filtering Stage**: Approximately 92% (460 questions) passed initial checks for syntax correctness, logical consistency, and content relevance.
- **Expert Validation Stage**: A team of programming educators rated the remaining 460 questions on three criteria: technical accuracy, pedagogical relevance, and linguistic clarity. Questions were scored on a 1–5 scale, and the results are summarized in Table 2.

Table 2. Expert Evaluation of Generated Questions (1–5 Scale)

Criteria	Average Score	Standard Deviation
Technical Accuracy	4.6	0.4
Pedagogical Relevance	4.3	0.5
Linguistic Clarity	4.2	0.6
Overall	**4.4**	**0.5**

These results indicate that the system successfully generates technically accurate and pedagogically sound questions, meeting instructional objectives.

5.2 Reinforcement Learning Adaptability

The reinforcement learning component was evaluated using simulated student profiles representing beginner, intermediate, and advanced learners. The system's ability to adapt question difficulty dynamically is shown in Fig. 6.

The adaptability of the RL agent in dynamically adjusting question difficulty based on simulated student performance over 20 iterations. For beginner profiles, the RL agent starts with 'Easy' questions and gradually increases difficulty to 'Medium' as performance improves, stabilizing by iteration 5. Intermediate profiles begin with 'Medium' difficulty questions and converge to 'Hard' difficulty by iteration 5, maintaining this level as performance stabilizes. Advanced profiles consistently receive 'Hard' questions from the outset, as no further adjustment is required.

5.3 System Efficiency

The system's efficiency was evaluated based on the time taken to generate, classify, and adapt questions. Table 3 summarizes the results.

The system achieved an average response time of 550 ms per question, making it suitable for real-time applications. The pre-caching of questions further ensured seamless operation in offline environments.

5.4 Diversity of Generated Questions

The diversity of generated questions was analyzed to ensure coverage across major programming concepts and difficulty levels. The results show:

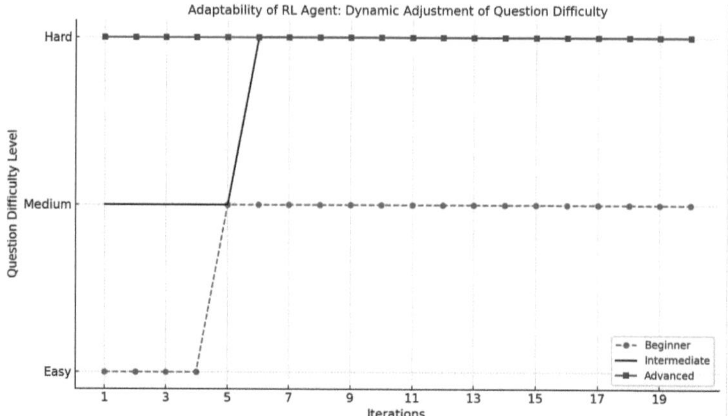

Fig. 6. Adaptability of RL Agent: Dynamic Adjustment of Question Difficulty.

Table 3. System Efficiency: Average Execution Times

Task	Execution Time (ms)
Question Generation (GPT)	280
Difficulty Classification (BERT)	120
RL-based Difficulty Adjustment	150
Total	**550 ms**

- Questions spanned core Java topics, including object-oriented programming, recursion, and data structures.
- Difficulty distribution was balanced, with 35% easy, 45% medium, and 20% hard questions.

These findings confirm that the system generates a diverse and representative set of questions that address a wide range of instructional needs.

5.5 Summary of Results

The evaluation demonstrates that the system effectively generates high-quality programming questions while dynamically adapting difficulty through reinforcement learning. Key highlights include:

- An expert-validated question quality score of **4.4** across all criteria.
- Efficient RL adaptability, with convergence achieved in **10 iterations**.
- A low execution time of **550 ms**, ensuring real-time performance.
- A balanced question diversity across topics and difficulty levels.

6 Discussion

The results demonstrate that the system successfully generates high-quality questions, achieving an average expert evaluation score of **4.4/5** across technical accuracy, pedagogical relevance, and linguistic clarity. Additionally, the RL agent effectively adapts question difficulty to simulated student performance, converging to appropriate levels within **10 iterations**. The system also achieves strong performance in terms of efficiency, with an average response time of **550 ms**, ensuring its suitability for real-time applications.

These findings highlight the system's potential as a scalable and robust solution for formative assessment in computer science education.

6.1 Illustrative Use Case

Consider a rural secondary school where electricity is intermittent, internet access is sporadic, and one teacher is responsible for 60 students with varying levels of programming experience. In such a setting, traditional digital learning platforms are infeasible. By deploying our system offline on low-cost laptops or local servers, students can engage with dynamic programming exercises that adapt to their performance in real time. A beginner student struggling with control flow concepts, for instance, receives simpler, scaffolded questions, while an advanced peer progresses to recursion and data structures. The teacher, relieved from manually tailoring content for each learner, focuses instead on coaching and addressing conceptual difficulties identified by the system's feedback analytics. In this context, our system not only supports differentiated instruction but also contributes to a more inclusive, effective, and dignified learning experience for students who might otherwise be left behind.

6.2 Challenges and Limitations

Despite the system's demonstrated strengths, it faces several challenges and limitations that merit further exploration.

Limitations of Pre-trained Models

While the use of pre-trained GPT and BERT models offers significant advantages in terms of scalability and accessibility, it also presents limitations. These include:

- **Programming-Specific Nuances:** Pre-trained models may occasionally misclassify question difficulty or generate ambiguous questions due to a lack of domain-specific fine-tuning.
- **Potential for Bias:** Since the models are trained on generic datasets, biases in the training data may affect the generated content.

Addressing these limitations in future iterations will involve fine-tuning the models on curated programming datasets, which can improve the alignment of generated questions with specific instructional goals while mitigating domain-specific errors.

Scalability and Computational Overhead
While offline functionality reduces reliance on connectivity, deploying the system in large-scale classroom settings may pose challenges. Running BERT and RL locally could strain devices with limited computational power. Exploring lightweight model alternatives or hybrid online-offline deployments could address these concerns.

Evaluation in Real-World Settings
The current evaluation relies on simulated learner profiles, which, while rigorous, may not capture the full range of behaviors and challenges encountered in real-world classrooms. Simulations provide controlled insights but lack the complexity of real-world interactions, such as diverse motivations, varied backgrounds, and unpredictable behaviors. Real-world validation is essential to capture these dynamics. Planned steps include:

- Pilot studies in small programming classes for initial testing.
- Longitudinal studies to evaluate engagement and knowledge retention.
- Iterative refinement based on feedback from students and educators.

6.3 Ethical Considerations

While the system is designed with accessibility and equity in mind, several ethical and practical challenges must be addressed to ensure responsible use. One key concern is the potential for algorithmic bias in question generation and difficulty classification. Since models like GPT and BERT are trained on large, general-purpose datasets, they may inadvertently reflect socio-cultural biases or use terminology that is inaccessible or inappropriate in some educational contexts. This risk is particularly critical when deploying AI tools in underrepresented or linguistically diverse communities. Ongoing bias audits, community-driven prompt tuning, and culturally localized datasets will be essential to mitigate this.

Additionally, the use of adaptive feedback systems in education raises questions of transparency and trust. Learners and educators may be unaware of how difficulty levels are determined or how performance data is used. To foster user trust and informed use, we plan to implement explainability features and clear consent workflows, especially for systems deployed in formal educational settings.

Data privacy and security also require careful management. In offline deployments, data may be stored locally on devices without centralized encryption or cloud safeguards. To address this, we propose integrating local encryption protocols and anonymized data logging, ensuring compliance with data protection regulations while retaining the system's offline functionality.

Finally, while reinforcement learning enables personalized experiences, over- personalization or poor reward tuning could inadvertently discourage learners by presenting them with consistently easy or hard content. To avoid this, a balanced reward framework and educator-configurable parameters will be included in future versions.

7 Conclusion

This work presents a transformative approach to personalized programming education, integrating GPT, BERT, and reinforcement learning (RL) into a cohesive, adaptive system. By enabling offline deployment, the system ensures accessibility, scalability, and equity, extending the reach of AI-driven assessments to underserved and resource-constrained regions. This innovative combination of cutting-edge AI technologies positions our system as a distinct and impactful solution for democratizing access to adaptive learning across diverse educational contexts. Future work will focus on fine-tuning models with domain-specific data, extending the system to other programming languages, and conducting longitudinal studies in real-world classroom settings to validate its broader applicability.

8 Future Work

A critical next step is evaluating the system in authentic educational settings. While simulation with synthetic learner profiles provides valuable insight into system adaptability, real-world testing is essential for assessing the practical impact on learners and educators. We plan to pilot the system in secondary schools and community learning centers in resource-constrained regions, beginning with a partnership network in Sub-Saharan Africa. The pilot will involve monitoring learning gains, student engagement, and usability feedback over a semester. Additionally, educators will be surveyed to assess workload reduction and the pedagogical relevance of system-generated content. This field study will provide empirical evidence of the system's social impact and inform iterative refinements to better serve diverse learners in marginalized contexts.

References

1. Luckin, R., Holmes, W., Griffiths, M., Forcier, L.B.: Intelligence Unleashed: An Argument for AI in Education. Pearson Education (2016)
2. VanLehn, K.: The relative effectiveness of human tutoring, intelligent tutoring systems, and other tutoring systems. Educ. Psychologist **46**(4), 197–221 (2011)
3. Folajimi, Y.: From gpt to bert: Benchmarking large language models for automated quiz generation. In: Proceedings of the 2024 ACM Virtual Global Computing Education Conference (SIGCSE Virtual 2024), pp. 312–313. ACM (2024)
4. Brown, T., Mann, B., Ryder, N., et al.: Language models are few-shot learners. In: Advances in Neural Information Processing Systems, vol. 33, pp. 1877–1901 (2020)
5. Corbett, A.T., Anderson, J.R.: Knowledge tracing: modeling the acquisition of procedural knowledge. User Model. User-Adap. Inter. **4**(4), 253–278 (1995)
6. Koedinger, K.R., Corbett, A.T., Perfetti, C.: The knowledge-learning-instruction framework: bridging the science-practice chasm to enhance robust student learning. Cogn. Sci. **36**(5), 757–798 (2012)
7. Doroudi, S., Aleven, V., Brunskill, E.: Where's the reward? a review of reinforcement learning for instructional sequencing. Int. J. Artif. Intell. Educ. **29**, 568–620 (2019)
8. Susnjak, T.: The current state and future prospects of ai-driven automated question generation. J. Educ. Technol. Soc. **25**(2), 44–57 (2022)

9. Buffardi, K., Edwards, S.H.: Codeworkout: Short programming exercises with automated guidance and assessment. In: Proceedings of the 2014 ACM Technical Symposium on Computer Science Education, pp. 123–128. ACM (2014)
10. Ihantola, P., Ahoniemi, T., Karavirta, V., Seppälä, O.: Autograder: Automated programming assessment tools. In: Proceedings of the 2015 ACM Conference on Innovation and Technology in Computer Science Education, pp. 21–26. ACM (2015)
11. Hou, X., Wu, Z., Wang, X., Ericson, B.J.: Codetailor: Llm-powered personalized parsons puzzles for engaging support while learning programming. In: Proceedings of the Eleventh ACM Conference on Learning@ Scale, pp. 51–62 (2024)
12. Ma, Q., Shen, H., Koedinger, K., Wu, S.T.: How to teach programming in the ai era? Using llms as a teachable agent for debugging. In: International Conference on Artificial Intelligence in Education, pp. 265–279. Springer, Cham (2024)
13. Matsuda, N., Cohen, W.W., Koedinger, K.R.: Applying reinforcement learning to educational practice: A case study of ai-driven dynamic assessment in math tutoring. In: Artificial Intelligence in Education Conference, pp. 249–260 (2017)

Understanding vs. Generation: LLMs Are Better at Comprehending Low-Resource Languages like Urdu Than Generating Text in Them

Taaha Saleem Bajwa$^{(\boxtimes)}$

Lahore, Pakistan
taaha.s.bajwa@gmail.com

Abstract. Large Language Models (LLMs) are predominantly trained on English data, leading to significant performance challenges for low-resource languages. This study focuses on Urdu as a case study to explore how LLMs process low-resource languages. We find that when prompted in a low-resource language, LLMs primarily reason internally in English, and this internal reasoning is more coherent than the generated text in the target language. This contrast highlights a gap between the model's ability to comprehend low-resource languages and its ability to generate text in them. By analyzing these mechanisms, this work underscores the need for targeted improvements to enhance LLM performance for low-resource languages such as Urdu.

Keywords: Natural Language Processing · Large Language Models · Low resource languages

1 Introduction

Most multilingual Large Language Models (LLMs) are trained on English-dominant corpora, which results in a significant performance gap across different languages. Furthermore, this multilingual capability is generally focused on high-resource languages such as French and German, leaving limited support for low-resource languages.

Previous studies have suggested that due to their English-centric training datasets, LLMs use English as their latent language even when prompted in another language [1]. The capability of LLMs to process non-English languages relies mostly on a very small number of neurons located in the final and initial layers [2], which are primarily involved in translating the prompt to and from the latent language.

By mechanistically removing the translation features in the final layer of the LLM, internal latent responses in English are obtained that are generally more coherent than the outputs generated by the LLM in the target language.

T. S. Bajwa—Independent.

Y. Folajimi et al. (Eds.): SIAI 2025, CCIS 2599, pp. 107–111, 2025.
https://doi.org/10.1007/978-3-031-98949-0_10

2 Objectives

This study aims to enhance our understanding of how LLMs process low-resource languages, using Urdu as a case study. Specifically, we investigate the extent to which LLMs can understand low-resource languages compared to their ability to generate coherent text in them.

This work seeks to deepen our understanding of multilingual processing in LLMs and contribute to future improvements in their ability to handle low-resource languages effectively.

3 Methodology

3.1 Dataset

For this study, a toy dataset of diverse questions in English was generated using ChatGPT [3]. These questions were then translated into urdu using Google Translate API. The selected questions are primarily those that require explanatory answers spanning several lines, as the focus of this study is to assess both the fluency and relevance of answers generated by LLMs in low-resource languages. Each question is provided in English along with a corresponding Urdu translation. The total dataset consists of 239 questions, of which 15 are used to identify translation features, while the remaining 223 are used for evaluation.

3.2 Models Used

Due to computational limitations, this study is conducted on smaller LLMs with 2 to 3 billion parameters. For this study, we use Gemma-2-2b [4] and Llama-3.2-3b [5].

3.3 Finding Translation Direction Across Layers

We find that for smaller LLMs, regardless of the task or prompt, low-resource language generation is primarily mediated by a single direction in the final layers, which can be easily removed mechanistically. To identify translation direction, we use a method inspired by [6]. The model is prompted with N Urdu and English questions separately, while caching their residual activations for each layer. After hit and trail, we select $N = 16$ for this study.

Let $R_{eng,l}$ represent the cached residual activations for English questions and $R_{urd,l}$ represent the cached residual activations for Urdu questions at layer l. This layer l is one of the final layers and upon hit and trial, its optimal value was found to be $l = 24$ for Gemma-2-2b and $l = 25$ for Llama-3.2-3b.

The mean residual activations for each language at layer l are computed as follows:

$$m_{eng,l} = \frac{1}{N} \sum_{i=1}^{N} R_{eng,l,i} \tag{1}$$

$$m_{urd,l} = \frac{1}{N} \sum_{i=1}^{N} R_{urd,l,i} \tag{2}$$

Next, the mean residual activations are subtracted, followed by normalization to find the translation direction for layer l:

$$d_{l,norm} = \frac{m_{urd,l} - m_{eng,l}}{\left|\left|m_{urd,l} - m_{eng,l}\right|\right|} \tag{3}$$

3.4 Removing Translation Direction

For each residual activation R_l *at layer l,* the translation direction is ablated by first computing its projection onto the direction $d_{l,norm}$, scaling it, and then subtracting the scaled projection:

$$R_{l,ablation} = R_l - \left(\left(R_l.d_{l,norm}\right).d_{l,norm}\right) \tag{4}$$

This procedure ablates the translation features.

4 Key Findings

LLM is prompted with different queries from the dataset and internal latent response (in English) and final response (in Urdu) are observed.

Figure 1 shows the percentage comparison of successful query understanding and text generation for Urdu. The results indicate that while both models exhibit some capability in understanding queries, their ability to generate coherent text in Urdu remains significantly lower. Llama-3.2-3b shows better performance than Gemma-2-2b in both tasks, but the gap between understanding and generation highlights the challenges LLMs face in low-resource language processing.

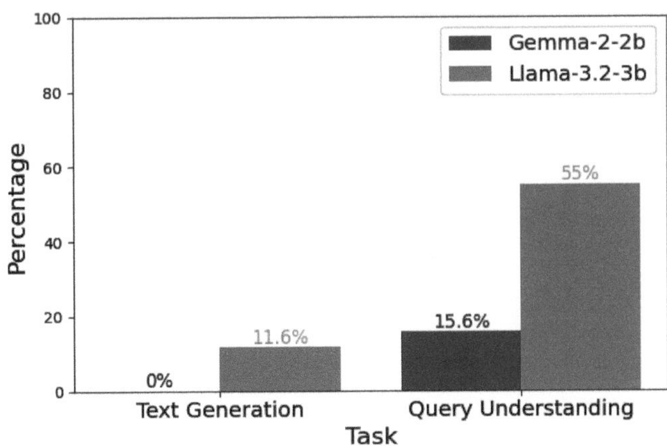

Fig. 1. Comparison of Percentage of successful text generation and successful query understanding in Urdu for Gemma-2-2b and Llama-3.2-3b.

5 Relevance to Emerging Regions

This study provides valuable insights into LLM behavior on low-resource languages, shedding light on the challenges these models face in processing and generating text in underrepresented languages. Urdu, a low-resource language, is spoken by 230 million people [7], yet receives minimal representation in LLM training data. This greatly restricts universal AI accessibility. Addressing these gaps is important to provide fairer AI access for people who speak different languages around the world.

6 Future Directions

This study focuses on Urdu as a case study, but its findings can be extended to other low-resource languages to further investigate how LLMs process and generate text in underrepresented languages. In the future, this work could serve as a benchmark for evaluating LLM performance on low-resource languages, providing a standardized way to measure and compare model capabilities.

Additionally, we plan to explore how these insights scale to larger LLMs, assessing whether increased model size improves language generation quality or if fundamental limitations persist. We also plan use these insights to build concrete methodologies to enhance LLM performance on low-resource languages, including improved training and inference strategies.

External Links.   Dataset, Code and Results can be found here - https://github.com/taaha/LRL_ processing

Appendix

The prompt used to generate the dataset can be seen in Table 1.

Table 1. Prompt used to generate dataset.

Prompt used
Generate one line diverse english questions that require explanatory answers

References

1. Wendler, C., Veselovsky, V., Monea, G., West, R.: Do llamas work in English? on the latent language of multilingual transformers. In: Ku, L., Martins, A., Srikumar, V. (eds.) ACL 2024, pp. 15366–15394. Association for Computational Linguistics, Bangkok (2024)
2. Tang, T., et al.: Language-specific neurons: the key to multilingual capabilities in large language models. In: Ku, L., Martins, A., Srikumar, V. (eds.) ACL 2024, pp. 5701–5715. Association for Computational Linguistics, Bangkok (2024)

3. ChatGPT. https://openai.com/chatgpt/overview/. Accessed 17 Oct 2024
4. Gemma Team, Riviere, M., Pathak, S., Sessa, P.G., Hardin, C., Bhupatiraju, S., et al.: Gemma 2: Improving Open Language Models at a Practical Size. arXiv preprint arXiv:2408.00118 (2024)
5. Dubey, A., Jauhri, A., Pandey, A., Kadian, A., Al-Dahle, A., Letman, A., et al.: The Llama 3 Herd of Models. arXiv preprint arXiv:2407.21783 (2024)
6. Arditi, A., Obeso, O.B., Syed, A., Paleka, D., Panickssery, N., Gurnee, W., Nanda, N.: Refusal in Language Models Is Mediated by a Single Direction. In: ICML 2024 Workshop on Mechanistic Interpretability (2024)
7. Hussain, C., Hussain, M.: Language politics in pakistan: urdu as official versus national lingua franca. Annal. Human Soc. Sci. **3**(2), 82–91 (2022)

Structured Reasoning for Fairness: A Multi-agent Approach to Bias Detection in Textual Data

Tianyi Huang[1,2](✉) and Elsa Fan[2]

[1] App Inventor Foundation, Cambridge, USA
`tianyi@appinventorfoundation.org`
[2] App-In Club, Fremont, USA
`elsa@appinclub.org`

Abstract. From disinformation spread by AI chatbots to AI recommendations that inadvertently reinforce stereotypes, textual bias poses a significant challenge to the trustworthiness of large language models (LLMs). In this paper, we propose a multi-agent framework that systematically identifies biases by disentangling each statement as fact or opinion, assigning a bias intensity score, and providing concise, factual justifications. Evaluated on 1,500 samples from the WikiNPOV dataset, the framework achieves 84.9% accuracy—an improvement of 13.0% over the zero-shot baseline—demonstrating the efficacy of explicitly modeling fact versus opinion prior to quantifying bias intensity. By combining enhanced detection accuracy with interpretable explanations, this approach sets a foundation for promoting fairness and accountability in modern language models.

Keywords: Bias Detection · Multi-Agent Systems · Trustworthy AI

1 Introduction

Words hold immense power in shaping perceptions, influencing social exchanges, and driving decision-making processes. In the era of large language models (LLMs), this power is amplified, as automated systems now participate in generating and interpreting large volumes of textual data at unprecedented scales [6, 24]. The reach of LLMs extends from assisting medical diagnoses and legal contract analysis to moderating online content and supporting educational tools [20]. Despite their remarkable influence and capabilities, these systems often inherit the biases embedded in their training data, risking the perpetuation of harmful stereotypes or discriminatory language [8, 17]. Equally problematic, subtle subjectivity and skewed phrasing may pass unnoticed, exposing end-users to outputs that inadvertently frame narratives in ways misaligned with fairness [3]. These challenges emphasize the urgent need for bias detection methods that not only identify problematic content but also clarify how and why biases arise [14] (Fig. 1).

T. Huang—Primary Author.

© The Author(s), under exclusive license to Springer Nature Switzerland AG 2025
Y. Folajimi et al. (Eds.): SIAI 2025, CCIS 2599, pp. 112–124, 2025.
https://doi.org/10.1007/978-3-031-98949-0_11

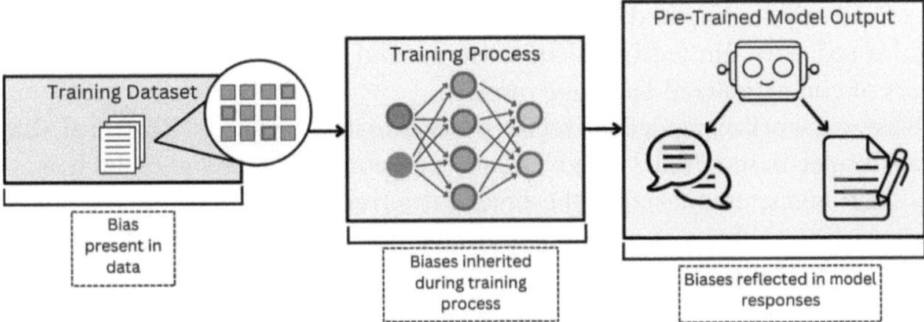

Fig. 1. An illustration of how biases present in a training dataset can be inherited by an AI model during training and reflected in the model's responses, potentially compromising objectivity.

Existing approaches to bias detection and mitigation often rely on static lexicons or predefined rules, which fail to capture the nuances of emerging or context- dependent biases [12, 25]. Another limitation is that these methods would lack explainability, reducing transparency in AI-driven decisions [22]. Moreover, certain methods simply mask biased terms or phrases without providing insights into the broader social or factual underpinnings of the bias [5]. Consequently, there remains a gap in the literature for frameworks that integrate factual verification, subjective analysis, and transparent explanations.

In this paper, we introduce a multi-agent framework that aims to tackle these shortfalls through systematically detecting bias and opinionated language in textual data through a structured reasoning process. Our pipeline includes:

1. A checker agent that classifies a statement as factual or opinion-based, removing ambiguity in subsequent analyses.
2. A validation agent that measures the intensity of bias in opinionated statements using a validity scoring mechanism, ensuring both subtle and overt biases are captured.
3. A justification module that provides explanations of the final classification, promoting better interpretability and transparency.

Beyond improving bias detection, our work contributes to broader efforts in creating accountable and socially responsible AI: by integrating this holistic approach, it holds promise for real-world deployments where unbiased AI outcomes are imperative—ultimately advancing the goal of developing AI technologies that uplift rather than undermine societal well-being.

2 Related Works

Research on bias in Natural Language Processing (NLP) has evolved considerably over the last decade, driven by growing concerns over computational models often reflecting and amplifying existing societal prejudices. Early approaches for addressing this issue largely focused on debiasing static word embeddings, as demonstrated by Bolukbasi et al. (NIPS 2016), who identified systematic gender biases in vector representations and proposed geometric alignment techniques to mitigate them [4]. While these initial efforts effectively highlighted the pervasiveness of stereotyping in word embeddings, they addressed only limited linguistic contexts and were insufficient in capturing the subtleties of contextualized language models.

Subsequent work expanded the focus to contextual embeddings. Zhao et al. illustrated how transformer-based models inadvertently perpetuate gender and racial biases across various NLP tasks, emphasizing the potential adverse consequences for downstream applications [26]. Efforts to measure and quantify bias in contextual representations often rely on carefully designed benchmarks and diagnostic tests, such as the StereoSet and CrowS-Pairs datasets, which reveal performance disparities correlated with sensitive attributes [18, 19]. However, purely quantitative evaluation methods can frequently overlook more nuanced forms of bias—particularly statements that embed subtle value judgments rather than including explicit biases [27].

A second line of inquiry examines explainability and interpretability as prerequisites for credible bias detection. Ribeiro et al. (KDD 2016) proposed Local Interpretable Model-Agnostic Explanations (LIME) to help end-users understand and trust classifier decisions [23]. Lundberg and Lee (NIPS 2017) introduced SHAP (SHapley Additive exPlanations), a unified framework for interpreting predictions across a variety of models [16]. While these techniques demystify model outputs by highlighting salient tokens or phrases, they do not always pin- point the origin of biases in training data or account for the degree to which an entire statement might be skewed.

More recent research ventures into multi-faceted bias detection that integrates social context, factuality checks, and user feedback loops. For instance, Field and Tsvetkov (2020) explored unsupervised methods for classifying gender in text, emphasizing the importance of considering contextual factors in bias detection [7]. In parallel, integrated pipelines for bias analysis—e.g., evaluating explicit sentiment, measuring harmful stereotypes, and quantifying subjectivity—have proven beneficial in tasks such as moderated content filtering and hate speech detection [2, 9, 15]. Still, many existing tools provide fragmented insights; some focus solely on word-level biases, while others rely on rigid rules that fail to adapt to evolving language trends. Moreover, transparency often remains insufficient: users may see a biased word flagged but lack an explanation grounded in factual evidence or logical reasoning.

Although bias detection in NLP has progressed through the advent of various approaches, current solutions fail to address three main problems: quantitative methods struggle in identifying nuanced biases, interpretability methods face difficulty in determining the full extent of bias in statements, and approaches involving integrated evaluation focus solely on subjective components. Consequently, these methods become ineffective in detecting statements that appear factual yet contain subtle biased language, as a deeper analysis rooted in factual evaluation and contextual understanding is required. Furthermore, these systems often fall short of producing valid and logical explanations for bias detection, hindering progress in ensuring full transparency for LLMs. To address these concerns, we propose a multi-agent framework that detects biases using a systematic approach. Unlike computational methods, our system determines implicit biases by distinguishing factual content from opinion-based text and quantifies varying degrees of bias to handle limitations present in the LIME and SHAP frameworks. Additionally, we also reduce the risks involved in analyzing subjectivity by focusing on classifying the nature of the statements themselves and evaluating their ability to be verified with evidence. These advances and the inclusion of justification responses place our framework as a solution for reinforcing fairness in AI systems (Fig. 2).

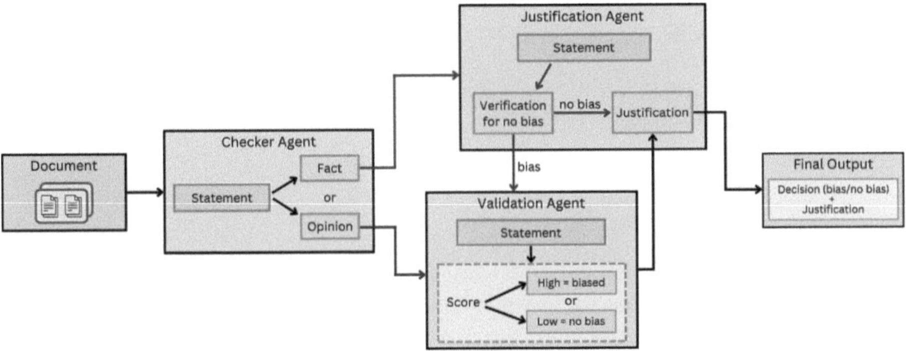

Fig. 2. Overview of the multi-agent bias detection pipeline. Text statements first enter a checker agent to be classified as *fact* or *opinion*. Factual statements are then verified by a justification agent for bias, while opinionated statements undergo evaluation by a validation agent. Finally, the system outputs a final decision (biased or unbiased) alongside a concise justification.

3 Methodology

3.1 Checker Agent: Fact vs. Opinion Classification

The initial step of our system is determining whether a statement is purely factual or contains subjective elements. We let S denote the statement and define a decision function:

$$Decision(S) = \begin{cases} \text{FACT,} & \text{if S is purely verifiable,} \\ \text{OPINION,} & \text{otherwise.} \end{cases}$$

A statement labeled as *FACT* is expected to be completely objective and testable against empirical evidence, while any presence of interpretive or persuasive language triggers an *OPINION* label. This initial filter ensures that the subsequent steps can be fitted to the specific nature of the statement.

3.2 Handling Factual Statements

If the checker agent outputs *FACT*, the pipeline applies a minimal bias verification step to confirm that the statement's language or presentation does not subtly introduce skew or partial framing. Specifically:

- **Factual-Bias Verification:** Factual statements may occasionally exhibit bias through wording, emphasis, or selective omission. If the pipeline detects no bias here, it forwards the statement to the justification writing with a "No Bias" outcome.

- **Potential Bias Escalation:** If this initial check suggests that the factual statement may contain bias, it is routed to the validation agent for a more in-depth inspection. This approach conserves computational costs by avoiding unnecessary full-scale analysis in obviously unbiased factual cases.

3.3 Validation Agent: Bias Scoring

All statements labeled *OPINION* by the checker agent, along with any factual statements flagged as potentially biased, are sent to the validation agent for full-scale analysis. Formally, we write:

$$Validate_{bias}(S) = f_{LLM}(S) \tag{1}$$

where f_{LLM} is a large language model tasked with assessing the extent of bias. For opinion statements, the agent looks for subjective or emotional language, strong value judgments, and imbalance in perspective. For escalated factual statements, it focuses on how factual content may be presented in a biased manner (e.g., emotive tone, selective emphasis). The validation agent then assigns a *Bias Level* of HIGH or LOW. We interpret HIGH as a binary predicted bias = True and LOW as predicted bias = False.

3.4 Justification Agent

Regardless of the validation outcome, a justification agent is called to generate a concise explanation of the verdict. This agent references the reasoning steps that led to either *No Bias* or *Bias*. Specifically,

- **No-Bias Cases:** The justification emphasizes the statement's objectivity and neutrality.
- **Biased Cases:** The justification pinpoints specific words, framing devices, or tones that caused the bias classification.

   This interpretability step aims to allow for greater transparency, especially in high-stake applications where reliability is paramount.

3.5 Final Output

After passing through the above stages, the pipeline produces two pieces of information:

1. **Binary Bias Classification:** "Bias" or "No Bias."
2. **Justification:** A concise explanation detailing the reasons behind the classification.

   These outputs are stored in a .json file for subsequent metric calculations and potential use in applications that require auditability or further inspections.

3.6 Implementation Details

We implement our pipeline in a modular structure by making asynchronous calls to a large language model at each agent stage:

– **Choice of LLM:** In our experiments, we primarily used *GPT-4o* to perform classification, bias verification, and justification generation [21]. Nonetheless, the design can be integrated with other LLMs as long as it follows the prompts and output formats.
– **Data Pipeline:** We randomly sample 1,500 labeled statements (biased and unbiased) from the WikiNPOV dataset, ensuring consistency via a fixed random seed [11]. Each statement travels asynchronously through the checker, bias-verification (if factual), validation, and justification steps, which support scalability in large datasets.

4   Baseline Approach

In addition to our multi-agent pipeline, we employ a *zero-shot baseline* for comparative evaluation. This baseline directly prompts GPT-4o (or any other preferred LLM) to classify each statement as either "biased" or "unbiased" without employing specialized fact-opinion segmentation [21]. The model operates with a single instruction focused solely on identifying bias in language or presentation, producing a one-word output per statement. This zero-shot approach serves as a valuable comparison for evaluating the pipeline's effectiveness in enhancing bias detection.

4.1   Performance Metrics

We measure the pipeline's effectiveness by comparing predicted labels ($\hat{y}$) against the ground truth ($\hat{y}$) on the sampled statements. Specifically, we compute the following:

We measure the effectiveness of both our multi-agent pipeline and the baseline by comparing their predicted labels $\hat{y}$ to the ground truth y for each of the sampled statements. Specifically, we use standard metrics such as accuracy, precision, recall, and F1 score:

$$\text{Accuracy} = \frac{\sum_{i=1}^{N} 1(\hat{y}_i = y_i)}{N}, \text{Pr ecision} = \frac{\sum_{i=1}^{N} 1(\hat{y}_i = 1 \wedge y_i = 1)}{\sum_{i=1}^{N} 1(\hat{y}_i = 1)},$$
$$\text{Recall} = \frac{\sum_{i=1}^{N} 1(\hat{y}_i = 1 \wedge y_i = 1)}{\sum_{i=1}^{N} 1(y_i = 1)}, F1 = 2 \cdot \frac{\text{Pr ecision} \cdot \text{Recall}}{\text{Pr ecision} + \text{Recall}} \tag{1}$$

5   Results

5.1   Comparing Pipeline and Baseline

We evaluate our *Pipeline* (checker–validation–justification) against a *Baseline* that relies on a single prompt to classify each statement as "biased" or "unbiased." Both methods use GPT-4o as their underlying LLM. Table 1 summarizes the results on the 1,500-statement sample from the WikiNPOV dataset, with performance metrics reported in percentages.

Table 1. Performance on a 1,500-statement subset of the WikiNPOV dataset (GPT-4o).

Method	Accuracy	Precision	Recall	F1 Score
Baseline	0.719	0.328	0.839	0.472
Pipeline	0.849	0.494	0.518	0.505

Statistical Significance. We conducted a two-proportion z-test to verify whether the accuracy difference between the two methods is statistically significant. Specifically, for the 1,500-statement test set, the Baseline correctly classifies approximately 1, 079 instances (71.9%), whereas the Pipeline correctly classifies around 1, 274 (84.9%). The resulting z-score is above 8.0, yielding a p-value below 10^{-6}. This indicates that our pipeline's improvement in accuracy is both practically meaningful and statistically robust.

Confusion Matrices. Figures 3 and 4 represent the confusion matrices for the Baseline and Pipeline, respectively. The Baseline demonstrates a relatively high Recall (few *false negatives*) but struggles with Precision, as evidenced by its numerous *false positives*. In contrast, our Pipeline achieves a more balanced distribution of true positives and true negatives, thereby attaining a higher F1 Score and offering superior overall reliability.

5.2   Adaptability to Different LLMs

To assess generality, we applied our Pipeline with three different state-of-the- art LLMs on a smaller 100-statement sample: GPT-4o, Claude 2.1, and Google

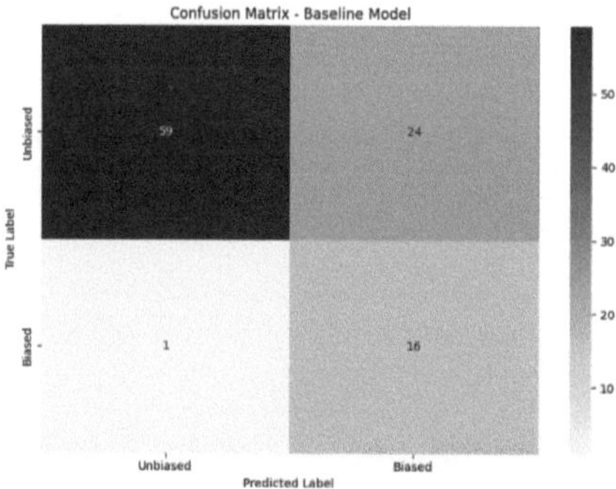

Fig. 3. Confusion Matrix for the Baseline on 100 WikiNPOV statements (GPT- 4o).

Gemini 1.5 Flash [1, 10, 21]. Table 2 shows that although there is some variability (especially in Recall) when switching to Claude, the Pipeline remains operational and achieves around 80% Accuracy in all settings. These findings show that our multi-agent design is not tightly coupled to one particular model and can be migrated to alternative LLMs.

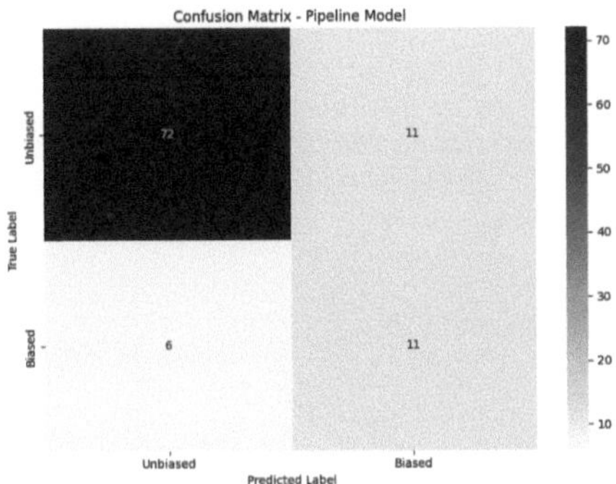

Fig. 4. Confusion Matrix for the Pipeline on 100 WikiNPOV statements (GPT- 4o).

5.3   Qualitative Insights

While the Baseline model allows for only a single-word prediction for each statement, our Pipeline provides a structured JSON output containing fields such as analysis, bias score, and justification. As shown in the listing below, the Baseline simply returns a label ("biased") for the example statement. Notably, the Pipeline not only labels a statement as *biased* or *unbiased* but also explains why it made that decision. This interpretability is beneficial for domains where traceability and reliability are important.

Table 2.   Performance of pipeline comparing three LLMs (100-statement subset), including GPT-4o, Claude 2.1, and Google Gemini 1.5 Flash.

LLM	Accuracy	Precision	Recall	F1 Score
GPT-4o	0.830	0.500	0.647	0.564
Claude	0.810	0.400	0.235	0.296
Gemini	0.780	0.696	0.780	0.736

```
 1   {
 2           "text": "A correct understanding...gravitation  ",
 3           "true_label": 0,
 4           "predicted_label": 0,
 5           "raw_response":  "biased"
 6   }    # Baseline
 7
 8   {
 9           "text": "A correct understanding...gravitation  ",
10           "true_label": 0,
11           "predicted_label": 0,
12           "statement_type": "opinion",
13           "analysis": {
14                   "fact_check": null,
15                   "bias_score":  high,
16                   "justification": "The statement shows bias by
             presenting general relativity as the only valid framework,
                which disregards alternative
             theories or interpretations. It lacks balance
             ..."
17           }} # Pipeline
```

5.4 Discussion

The significant improvements in accuracy ($p < 10^{-6}$) validate the effectiveness of this multi-agent pipeline for bias detection tasks with a checker, validation, and justification agent. Additionally, the framework's adaptable architecture allows the LLM model to be switched, making it extensible to new and emerging models.

Here, N is the total number of evaluated statements, and $1(\cdot)$ is the indicator function. These metrics, complemented by detailed logs of final decisions (e.g., statement type, bias level, justification), enable full assessments of our methodology's robustness.

6 Limitations and Future Work

6.1 Limitations

Although our framework shows promise in improving bias detection for textual data, several challenges should be considered:

- **Dataset Dependency:** The WikiNPOV dataset that our system is evaluated on consists of limited data that, while comprehensive, may not encompass all biases present in real-world situations [11]. The framework's performance could vary when applied to domains or datasets with different patterns or contextual requirements, drawing the need for broader dataset evaluations.
- **Uncertainty in Classifying Facts and Opinions:** Our framework relies on a binary categorization of statements as either factual or opinion-based. Statements blending factual and opinionated elements may lead to misclassifications, reducing the accuracy of subsequent bias detection. This limitation underscores the need for a more nuanced classification mechanism capable of handling hybrid statements.
- **Missed Insights in Bias Intensity:** The system currently outputs a binary bias intensity score (High or Low), which may oversimplify the complexity of opinionated statements. This approach could overlook the nuances of bias severity, limiting the depth of insights provided by the system and its capacity for complex justifications.

6.2 Future Work

Our research presents several paths where future work can be introduced to improve the effectiveness and applicability of our bias detection framework:

- **Specific bias evaluation:** To introduce a more comprehensive detection system, future work could focus on producing a more detailed assessment of biased statements and providing better explainability. For example, a bias determiner agent could be implemented in the framework to classify for specific social or cultural biases rather than solely distinguishing between biased and unbiased text.
- **Percentage scoring for bias:** Transitioning from binary bias scores to a continuous or percentage-based scale would enable a more detailed assessment of bias intensity. This change could provide users with richer interpretability and a clearer understanding of the system's decision-making process.

- **Contextual bias detection:** Incorporating context-aware models or Retrieval-augmented Generation (RAG) systems could improve the framework's ability to detect context-specific biases [13]. These enhancements would allow for more detailed justifications grounded in relevant contextual information.
- **Broader dataset application:** To improve generalizability, the framework should be evaluated on datasets from diverse domains such as medicine, law, and journalism, where objectivity and fairness are critical. This would help assess the system's adaptability to varying linguistic and contextual nuances.

7 Ethical Considerations

The development of our bias-detection framework requires careful attention to transparency, fairness, and the limitations inherent in bias evaluation. As biases are often subjective and embedded in the training data, our framework's accuracy depends on the diversity and representativeness of the data sources used. Ensuring inclusivity in training data and incorporating broader contextual information are necessary steps to address this challenge. Additionally, while the framework provides justifications for its classifications, these explanations may not always fully capture the nuances of complex biases, highlighting the need for more detailed and comprehensive rationalizations. By addressing these concerns, we aim to create a more equitable and transparent system that builds trust and supports ethical AI development.

8 Conclusion

This paper introduced a multi-agent framework for bias detection that combines fact–opinion classification, bias verification, and concise justification. Across 1,500 statements from the WikiNPOV dataset, the framework significantly out-performed a zero-shot baseline ($p < 10^{-6}$), demonstrating the impact of a structured, agent-based pipeline for improving both accuracy and interpretability. Additionally, experiments with multiple LLMs, including GPT-4o and Claude 2.1, confirmed the system's adaptability. By generating transparent justifications for each classification, our method offers practical advantages in domains requiring trustworthy AI-driven decisions. Future directions include expanding the pipeline to multilingual contexts, integrating external knowledge bases, and adding confidence calibration to further bolster reliability and user trust. Ultimately, this framework contributes to AI-driven systems aiming to promote fairness and accountability, where structured, transparent bias detection can function as a protection in the broader pursuit of trustworthy AI.

References

1. Anthropic (2023) Introducing claude 2.1. https://www.anthropic.com/news/claude-2-1
2. Aoyagui, P.A., Ferguson, S., Kuzminykh, A.: Exploring subjectivity for more human-centric assessment of social biases in large language models (2024). https://arxiv.org/abs/2405.11048

3. Bender, E.M., Gebru, T., McMillan-Major, A., Shmitchell, S.: On the dangers of stochastic parrots: Can language models be too big? In: Proceedings of the 2021 ACM Conference on Fairness, Accountability, and Transparency, Association for Computing Machinery, New York, NY, USA (2021). https://doi.org/10.1145/3442188.3445922
4. Bolukbasi, T., Chang, K.W., Zou, J., Saligrama, V., Kalai, A.: Man is to computer programmer as woman is to homemaker? debiasing word embeddings (2016). https://arxiv.org/abs/1607.06520
5. Dev, S., Li, T., Phillips, J., Srikumar, V.: On measuring and mitigating biased inferences of word embeddings (2019). https://arxiv.org/abs/1908.09369
6. Devlin, J., Chang, M.W., Lee, K., Toutanova, K.: Bert: Pre-training of deep bidirectional transformers for language understanding (2019). https://arxiv.org/abs/1810.04805
7. Field, A., Tsvetkov, Y.: Unsupervised discovery of implicit gender bias (2020). https://arxiv.org/abs/2004.08361
8. Gallegos, I.O., et al.: Bias and fairness in large language models: A survey (2024). https://arxiv.org/abs/2309.00770
9. Garg, A., Srivastava, D., Xu, Z., Huang, L.: Identifying and measuring token-level sentiment bias in pre-trained language models with prompts (2022). https://arxiv.org/abs/2204.07289
10. Google (2024) Gemini 1.5 flash-8b is now productionready.URLhttps://developers.google blog.com/en/gemini-15-flash-8b-is-now-generally-available-for-use/
11. Hube, C., Fetahu, B.: Neural based statement classification for biased language. In: Proceedings of the Twelfth ACM International Conference on Web Search and Data Mining, ACM, WSDM '19, pp. 195–203 (2019). https://doi.org/10.1145/3289600.3291018
12. Husse, S., Spitz, A.: Mind your bias: A critical review of bias detection methods for contextual language models (2022). https://arxiv.org/abs/2211.08461
13. Lewis, P., et al.: Retrieval- augmented generation for knowledge-intensive nlp tasks (2021). https://arxiv.org/abs/2005.11401
14. Li, Y., Du, M., Song, R., Wang, X., Wang, Y.: A survey on fairness in large language models (2024). https://arxiv.org/abs/2308.10149
15. Liu, Y., Yang, K., Qi, Z., Liu, X., Yu, Y., Zhai, C.: Prejudice and volatility: A statistical framework for measuring social discrimination in large language models (2024). https://arxiv.org/abs/2402.15481
16. Lundberg, S., Lee, S.I.: A unified approach to interpreting model predictions (2017). https://arxiv.org/abs/1705.07874
17. Mei, K., Fereidooni, S., Caliskan, A.: Bias against 93 stigmatized groups in masked language models and downstream sentiment classification tasks. Proceedings of the 2023 ACM Conference on Fairness, Accountability, and Transparency (2023). https://api.semanticscholar.org/CorpusID:259129801
18. Nadeem, M., Bethke, A., Reddy, S.: Stereoset: Measuring stereotypical bias in pretrained language models (2020). https://arxiv.org/abs/2004.09456
19. Nangia, N., Vania, C., Bhalerao, R., Bowman, S.R.: CrowS-pairs: A challenge dataset for measuring social biases in masked language models. In: Proceedings of the 2020 Conference on Empirical Methods in Natural Language Processing (EMNLP), Association for Computational Linguistics, Online (2020). https://aclanthology.org/2020.emnlp-main.154/
20. Omar, M., Nadkarni, G.N., Klang, E., Glicksberg, B.S.: Large language models in medicine: a review of current clinical trials across healthcare applications. PLOS Digital Health 3 (2024). https://doi.org/10.1371/journal.pdig.0000662
21. OpenAI (2024) Hello gpt-4o. https://openai.com/index/hello-gpt-4o/
22. Petkovic, D.: It is not "accuracy vs. explainability" – we need both for trustworthy ai systems (2022). https://arxiv.org/abs/2212.11136
23. Ribeiro, M.T., Singh, S., Guestrin, C.: "Why should i trust you?": Explaining the predictions of any classifier (2016). https://arxiv.org/abs/1602.04938

24. Vaswani, A., et al.: Attention is all you need (2023). https://arxiv.org/abs/1706.03762
25. Webster, K., et al.: Measuring and reducing gendered correlations in pre-trained models (2021). https://arxiv.org/abs/2010.06032
26. Zhao, J., Wang, T., Yatskar, M., Cotterell, R., Ordonez, V., Chang, K.W.: Gender bias in contextualized word embeddings (2019). https://arxiv.org/abs/1904.03310
27. Zhao, Y., Wang, B., Wang, Y.: Explicit vs. implicit: Investigating social bias in large language models through self-reflection (2025). https://arxiv.org/abs/2501.02295

Regional Determinants of Domestic Violence in India: Machine Learning Approach

Mansi Sharma[(✉)] and Greeshma Balabhadra[(✉)]

Stony Brook University, Stony Brook, NY 11794, USA
{mansi.sharma,greeshma.balabhadra}@stonybrook.edu

Abstract. In this paper, we analyze the factors affecting the likelihood of an Indian woman experiencing domestic violence. In addition, we also analyze how these factors vary across six regions in India. Using a nationally representative survey, we employ machine learning (ML) techniques such as logistic regression, Least Absolute Shrinkage and Selection Operator (LASSO), Random Forest (RF), Extreme Gradient Boosting (XGBoost), and SHapley Additive exPlanations (SHAP), along with dimension reduction methods like autoencoder. Among logistic regression, LASSO, and XGBoost, RF consistently outperforms the others in both the full sample and regional analyses. While different models prioritize features differently, key predictors of spousal abuse — husband's control issues, woman's age at first cohabitation, woman's family background, and her physical stature — consistently emerge as significant both in the overall sample and in the subsamples divided across the six regions in India. However, at the regional level, cluster analysis reveals significant intra-regional variation, reinforcing the need for localized interventions to effectively address domestic violence in India.

Keywords: Domestic Violence · Machine Learning · Clustering

1 Introduction

The detrimental effects of domestic violence on a victim's overall well-being are increasingly recognized [2].[1] A victim of domestic violence often suffers from physical and emotional pain, mental health issues, reduced productivity at work, alcohol abuse, unplanned pregnancies, miscarriages, maternal morbidity, stillbirth, and forced abortions [3–5, 9–11]. Beyond the immediate impact, domestic violence has long-term repercussions for future generations. For example, the United Nations (UN) reports that a child who witnesses violence at home is more likely to perform poorly in school and faces a higher risk of becoming a victim or a perpetrator of violence in adulthood [6].

In India, the country of study, traditional norms—such as the belief that *"beating is a husband's right and a wife's due"*—along with conservative practices and early

---

M. Sharma and G. Balabhadra—Contributed equally to this work.

[1] We use *spousal abuse, intimate partner violence, and domestic violence* interchangeably throughout this paper.

© The Author(s), under exclusive license to Springer Nature Switzerland AG 2025
Y. Folajimi et al. (Eds.): SIAI 2025, CCIS 2599, pp. 125–141, 2025.
https://doi.org/10.1007/978-3-031-98949-0_12

marriage, increase women's vulnerability to spousal abuse [9, 11, 22]. According to the most current National Family Health Survey data (NFHS-5), approximately 31% of ever-married Indian women have experienced domestic violence in their lifetimes.

Given this context, we study the determinants of domestic violence and how they vary across different regions in India. In particular, we present the top ten predictors of domestic violence in all of India and the top five predictors across different regions. We present regional-level predictors because cultural norms and practices differ by geography, which may exacerbate or mitigate domestic violence [15, 23, 24].[2] In addition, the most significant contributors to domestic violence in one region may not hold the same weight elsewhere. By conducting a regional analysis, we argue that targeted, group-specific interventions are more effective in addressing domestic violence than a one-size-fits-all approach.

We build on the approach of [20] and adopt a machine-learning framework to analyze the factors contributing to domestic violence. However, unlike [20], we also examine how these factors vary across regions using ML techniques. We employ various methods; for example, we start with logistic regression and extend to LASSO, Random Forest (RF), and gradient-boosted decision trees (XGBoost) to identify key features. These models capture linear and complex non-linear relationships between socioeconomic, demographic, and behavioral variables. To measure the prediction performance of different ML models, we use various metrics—Precision, Recall, Accuracy, and F1-score—that reflect the sensitivity of the problem, as our dataset is imbalanced due to the underreporting of spousal abuse. Compared with the traditional logistic regression, RF and XGBoost have higher prediction accuracy.

In addition, relative to [20], we also report SHAP (SHapley Additive exPlanations) values [43], which provide interpretable insights into feature importance both globally and locally. To further capture the structural dynamics of these features across regions, we apply an autoencoder for dimensionality reduction followed by K-means clustering. This analysis highlights intra-regional variation and reveals how each region is distributed across clusters based on top predictive features.

We apply machine learning (ML) techniques to study the factors affecting the risk of domestic violence because [20, 37, 40] argue that ML methods are more efficient in uncovering the predictors of spousal abuse and are more accurate in predicting instances of spousal abuse compared with the conditional probability model, logistic regression.

Beyond our comparison with [20], our study also differs from [41], who use tree-based ML models and logistic regression to examine spousal abuse determinants among 1,816 South African women. While [41] focuses on South Africa, our study is based on India. Within the Indian context, [12] apply ML models, specifically neural networks, to investigate factors influencing marital sexual violence.

To our knowledge, this is the first study to use ML to analyze regional variations in domestic violence experiences across India and to show that different regions may share some common clusters. That is, even if North and South India differ geographically, within these regions, there may be groups of women who share similar characteristics and are equally likely to experience or not experience spousal abuse, as demonstrated by the cluster analysis.

---

[2] Refer to Table 4 to see how India is divided into six regions.

We also contribute to the literature on the determinants of spousal abuse. For example, [13] model violence as a tool used by a husband to control his wife, while a wife may tolerate it for financial security. [33, 34], and [32] show that economic empowerment lowers abuse risk, while [31] find empowerment programs reduce reported abuse. [11] show that delaying marriage decreases the risk of spousal abuse, as an older bride tends to be more educated and marry a better-educated man. Education delays marriage [18].

Building on these broader insights, we now turn to the Indian context, where several factors have been identified as key predictors of spousal abuse. In India, key abuse predictors include education, caste, intergenerational transmission of violence, husband's controlling behavior, substance abuse, and household wealth [7, 19, 26, 29]. Furthermore, [30] find that longer marriages and childlessness increase abuse risk; however, [34] argue that marriage stability achieved through longer duration reduces it. In addition, [17] find that childbirth delays the onset of abuse, linking birth to the husband's satisfaction within the marriage.

We find that different ML models prioritize features differently. Key predictors of spousal abuse—husband's control issues, woman's family background, and her physical stature—consistently emerge as significant both in the overall sample and in the subsamples divided across the six regions in India. However, at the regional level, from cluster analysis, we find significant intra-regional variation, reinforcing the need for localized interventions to address domestic violence in India effectively.

The remainder of this paper is written in the following way. Section 2 describes the data and outlines the sample selection criteria used for the analysis. Section 3 details the various machine learning models and the estimation methodology applied to examine the determinants of spousal abuse. Section 4 presents the main empirical results at national and regional levels. Finally, Section 5 concludes with key policy recommendations and guidance for future work.

2 Data and Sample Selection

For our analysis, we use the fifth round of the National Family Health Survey (NFHS–5). It is a repetitive cross-sectional and nationally representative survey undertaken across all Indian states and union territories. Note that NFHS–5 is part of the Demographic and Health Surveys (DHS) program, which is conducted in over 90 countries across the world. As a result, we can replicate the study and do a comparative analysis across different countries [1].

The survey includes women aged 15 to 49 and men aged 15 to 54, and collects information on a wide range of topics. These topics include whether a woman has experienced any form of physical or sexual violence by a current or former partner, as well as issues related to fertility outcomes, family planning and contraceptive use, water and sanitation, lifestyle, attitudes toward spousal abuse, age at marriage and menstruation, healthcare utilization, and nutrition. The NFHS-5 reports information about the prevalence of violence (both over the lifetime and in the past 12 months) on the sub-sample of women selected for the domestic violence module.

We use the couple dataset, which provides information on married or cohabiting men and women. Table 1 shows the sample selection criteria for our study. We start with a sample of 57,693 and exclude couples for various reasons, as listed in Table 1, resulting in a final sample of 45,361. We exclude couples where the wife is not selected for the domestic violence module, which is conducted on a subsample of women. Second, we exclude couples where either caste information or the wife's height is missing.

Table 1. Study Sample Selection

Causes for Exclusion	#Obs Lost	#Remaining
Couples' Dataset		
Couples in the survey		57,693
Not selected for the Domestic Violence Module	11,205	46,488
Don't know response to the caste	253	46,235
Missing information on wife's height	874	45,361

For our analysis, domestic violence is a binary indicator that takes a value of one if a woman has experienced any form of physical (pushed, slapped, kicked, strangled, threatened with a knife or gun), emotional, or sexual violence by her husband/partner. Table 2 shows that 26% of the women in the sample have experienced less severe forms of violence, such as being pushed, slapped, punched, or having their arms twisted. Meanwhile, 30% of the women have experienced some form of violence in their lifetime from a partner or husband. It is important to note that these figures are likely under-reported due to stigma, fear, and the threat of retaliation [16].

Table 2. Dependent Variables

Types of Violence	Mean	Std. Dev.
Experienced any emotional violence	0.12	0.32
Experienced any less severe violence	0.26	0.44
Experienced any severe violence	0.07	0.26
Experienced any sexual violence	0.05	0.22
Experienced any form of violence	0.30	0.46

Following [41] and [38], we incorporate various individual and household-level factors as explanatory variables. These include women's current age, age at first cohabitation, height, body mass index (BMI), gender of the household head, age of the household head, religion, caste, educational attainment of both husband and wife, wealth status, number of living children, number of household members, number of sons at home, number of daughters at home, marital control face by a woman, partner's alcohol usage, history of abuse, place of residence (rural or urban), family background of both husband and wife, and attitudes toward spousal abuse.

3 Model and Estimation

3.1 Preprocessing of Data

For our analysis, we select 74 variables from the couples' dataset, in line with [41] and [38], a mix of boolean, numeric, and categorical. We combine the variables that can be interpreted as proxies for outcome variables into a single variable. For example, instead of using the three separate measures of violence, we create a binary indicator that takes the value of 1 if a woman has experienced any form of violence. Similarly, instead of using six separate variables to measure marital control faced by a woman, we create an aggregate index by summing all the binary responses.

In addition, to capture attitudes toward wife-beating, we create a binary indicator that takes the value of 1 if a respondent justifies spousal abuse in any of the seven predefined instances. This helps consolidate attitudinal information into a single interpretable feature.

Since domestic violence cases are underreported, the dataset is imbalanced, with fewer positive (i.e. reported violence) cases than negative cases. This imbalance can bias the model performance, especially for the classification tasks. To address this, we use the resampling strategy, Synthetic Minority Oversampling Technique (SMOTE), to generate synthetic data points for the minority class to balance class distribution. In addition, for our ML models, we incorporate class weighting strategies, setting *class weight = 'balanced'* in RF and tuning *scale pos weight* in XGBoost to ensure the minority class receives enough attention during data training.

For the regression models, we use min-max scaling for all the numerical (continuous) variables. Next, with regard to the categorical variables, we compare the distribution of the variable to our dependent variable and combine some sparse and similar categories to improve the interpretability of the model. Finally, we use *One-Hot encoding* to transform categorical variables into binary indicator variables, ensuring compatibility with ML algorithms that require numerical input.

3.2 Machine Learning Models

We use logistic regression as a benchmark and tune LASSO, Random Forest, and gradient-boosted decision trees (XGBoost) for feature selection and prediction analysis. We select these models for their strengths in handling our dataset's behavioral, socioeconomic, and demographic features and imbalanced nature. Feature selection is critical to building an efficient, interpretable model. Focusing on the most relevant predictors helps reduce dimensionality, improve computational efficiency, and enhance model generalization. In addition, it allows us to identify the most important determinants of spousal abuse, which policies can target.

Logistic Regression: Logistic regression is a simple, interpretable model used as a benchmark in our analysis. The mathematical formulation is

$$P(y = 1|X) = \frac{e^{X^T \beta}}{1 + e^{X^T \beta}}$$

where y is a binary indicator that takes the value 1 if a woman has experienced any form of violence and 0 otherwise, X is a vector of exogenous variables (including a constant term), and β is a vector of parameters we estimate using maximum likelihood estimation.

LASSO: The dataset contains numerous behavioral, socioeconomic, and demographic features; therefore, it is important to narrow down the key predictors. We use the Least Absolute Shrinkage and Selection Operator (LASSO), which applies l_1 regularization and simplifies the model by selecting only the most relevant features. LASSO penalizes the absolute magnitude of the coefficients, forcing some of them to become exactly zero, thereby eliminating irrelevant or less significant predictors. This results in intrinsic feature selection during training, enhancing model interpretability. The logistic regression model with LASSO regularization aims to minimize the following objective function:

$$min_\beta \left[-\frac{1}{N}\sum\nolimits_{i=1}^{N} (y_i log p_i + (1 - y_i)log(1 - p_i)) + \lambda \sum\nolimits_{j=1}^{p} |\beta_j| \right]$$

where $p_i = \frac{e^{X_i^T \beta}}{1+e^{X_i^T \beta}}$, and λ is the regularization parameter controlling the sparsity. LASSO effectively removes irrelevant predictors, improving interpretability without losing predictive performance.

Random Forest: Random Forest (RF) is an ensemble method that builds multiple decision trees using bootstrap samples and random feature subsets. It captures complex nonlinear relationships and interactions between features while maintaining robustness to overfitting. Feature importance is computed based on the average reduction in impurity across trees, providing insights into the most influential predictors. It also provides features selection and ranks them based on their overall contribution to the model. It is particularly useful for handling mixed-type data and noisy features.

Gradient Boosted Decision Trees: Extreme Gradient Boosting (XGBoost) is a highly efficient gradient-boosted decision tree algorithm optimized for speed and accuracy. It builds trees sequentially, and each new tree corrects the residuals of the precious ensemble. XGBoost supports custom loss functions and class weights, making it well-suited for imbalanced data sets. It also captures subtle patterns in the data by iterative improvements in prediction performance. XGBoost calculates the feature importance based on metrics like gain (improvement in accuracy in splitting on a feature), cover (frequency of the feature usage), and weight (number of times a feature is used in splits). It handles high-dimensional and imbalanced datasets, providing better insights into feature relevance.

SHAP (SHapley Additive exPlanations): SHAP is another tool based on cooperative game theory and is used to interpret predictions from complex models like RF and XGBoost. It provides a global understanding of how each feature contributes more across the dataset and a local understanding of how specific features influence individual predictions. We include SHAP summary plots to visualize the direction (positive or negative) and the magnitude of the feature effects, enhancing transparency and trust in model decisions.

3.3 Training and Hyperparameter Tuning

Each model is trained and tuned separately using 75% of the dataset for training and 25% for testing out-of-sample performance. To maintain class distribution in the imbalance dataset, we use stratified sampling during train-test split and stratified k-fold cross-validation (with $k = 5$) during hyperparameter tuning. This helps reduce overfitting, ensures generalizability to unseen data, and yields stable performance metrics.

We perform grid search cross-validation using the ROC-AUC score as the primary metric due to its threshold-independent nature and ability to handle class imbalance effectively. For additional insight into the model performance, we also compute precision, recall, and F1-Score, which are critical in imbalanced class distributions.

For RF model, we tune the following hyperparameters

- n_estimators : [100, 150, ..., 500]
- max_depth : [5, 10, ..., 20]
- min_samples_split: [2, 4, ..., 10]

For XGBoost model, we tune

- n_estimators: [100, 200, 300]
- learning_rate: [0.01, 0.05, 0.1]
- max_depth: [3, 6, 10]
- scale_pos_weight: tuned based on class imbalance ratio

For LASSO, we tune the regularization parameter λ using cross-validation, selecting the value that minimizes log-loss.

To ensure reproducibility, a fixed random seed is used throughout the training and cross-validation. Once optimal hyperparameters are selected, the final model is retrained on the whole training set and evaluated on the held-out test set.

Finally, we compare all models using a consistent set of metrics and interpret feature importance using SHAP values (for RF and XGBoost) and non-zero coefficients (for LASSO), with an emphasis on predictive accuracy and interpretability.

3.4 Evaluation Metrics

While evaluating different ML models, we consider various metrics that reflect the sensitivity of the problem as the data is imbalanced. The data is significantly skewed due to underreporting of domestic violence, which makes conventional accuracy based evaluation insufficient. Therefore, we focus on Precision, Recall, False Positive Rate (FPR), Accuracy, F1-Score, and AUC-PR as primary metrics.

- **Precision** measures the proportion of the true positive predictions among all the positive predictions. High precision means fewer false alarms, which is important to avoid unnecessary interventions or misclassifications - especially important in resource-constrained or sensitive policy settings.
- **Recall** measures how many actual positive cases are correctly identified, which is critical, as missing true cases of domestic violence can have serious consequences.
- **Accuracy** reflects the overall correctness of a model by calculating the proportion of correctly classified instances out of all predictions.

- **F1-Score** is the harmonic mean of precision and recall, balancing the trade-off between both metrics. It is especially useful when there is an uneven class distribution, and we want to capture both types of errors.
- **False Positive Rate (FPR)** is the proportion of actual negative cases incorrectly classified as positive. A lower FPR indicates fewer incorrect positive predictions, which is particularly important in sensitive applications.
- **AUC-PR** (Area Under the Precision-Recall Curve) is better suited than ROC curves for imbalanced data, as it emphasizes performance on the positive class and avoids misleading interpretations when true negatives dominate.
- **ROC-AUC** (Receiver Operating Characteristic – Area Under the Curve) measures how well a model distinguishes between classes by plotting the True Positive Rate (Recall) against the False Positive Rate (FPR) at various thresholds. A higher AUC (closer to 1) indicates better class separation, while a value of 0.5 suggests random guessing, making it a crucial metric for evaluating models on imbalanced datasets. Given a confusion matrix, we define metrics as follows:

$$Precision = \frac{TP}{TP + FP}$$

$$Recall = \frac{TP}{TP + FN}$$

$$FPR = \frac{FP}{FP + TN}$$

$$Accuracy = \frac{TP + TN}{TP + TN + FP + FN}$$

$$F1Score = 2 * \frac{Precision * Recall}{Precision + Recall}$$

where TP is true positive, FP is false positive, TN is true negative, and FN is false negative.

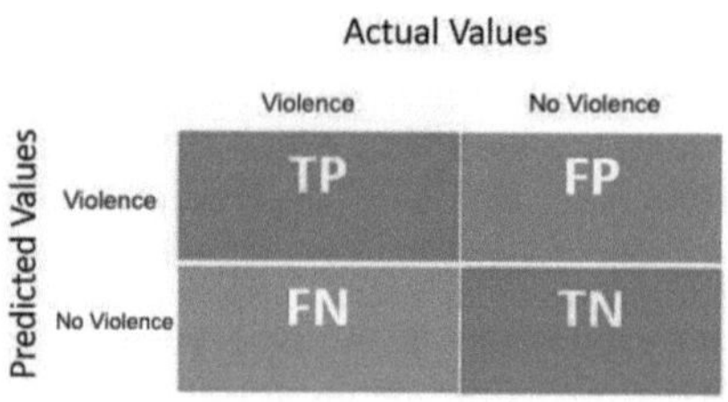

Fig. 1. Confusion matrix. Here, TP is true positive, FP is false positive, FN is false negative, and TN a is true negative

3.5 Clustering and Dimension Reduction

We perform clustering analysis to uncover latent group structures within the data, enabling segmentation based on a wide range of behavioral, socioeconomic, and demographic features. Starting with 74 variables, One-Hot encoding of categorical features expands the dataset to 274 variables. To handle this high dimensional space and improve the clustering quality, we first normalize all the features to [0, 1] and then apply an autoencoder for dimensionality reduction.

Autoencoders are neural network based architectures well suited for capturing nonlinear relationships across the features. They compress the high-dimensional data into low-dimensional latent space by learning the most informative representations by minimizing reconstruction loss. This approach reduces noise and also enhances the clustering process by projecting the data into a more compact and structured manifold. We use a symmetric autoencoder with ReLU activations and a bottleneck layer of size k(tuned via validation loss). trained using mean squared error as the loss function.

Following dimensionality reduction, we apply the MiniBatch K-Means algorithm—a scalable variant of traditional K-Means designed for large datasets. It improves computational speed by processing random mini-batches of the data at each iteration. To determine the optimal number of clusters, we use the elbow method, plotting the within-cluster sum of squares (inertia) against varying cluster counts. The *"elbow point"*—where the rate of decrease in inertia levels off—indicates the ideal number of clusters.

Finally, we interpret and profile the resulting clusters by analysing the mean feature values within each cluster. This helps identify the dominant socio-demographic or behavioral characteristics that define each segment and offer actionable insights for policy and intervention strategies.

4 Empirical Results

4.1 National Level Analysis

We compare logistic regression, LASSO, RF, and XGBoost based on precision, recall, F1-score, and AUC-PR, as shown in Table 3. Logistic regression shows moderate precision but performs poorly on recall, indicating a higher number of false negatives. LASSO offers marginal improvements over logistic regression, particularly in recall and AUC-PR. XGBoost outperforms RF in terms of precision, suggesting it generates fewer false positives. However, RF achieves the best balance across all metrics, with the highest F1-score, making it the overall best-performing model.

Considering the performance of RF and XGBoost in Table 3, we use both models for feature selection. The top 10 ranked features in the national-level analysis and their respective importance scores for both models are presented in Figs. 2 and 3. While we present the top features identified by both models, for interpretation, we focus on the top features from RF, given its superior performance in Table 3. We find that the husband's control issues emerge as the most significant factor associated with the wife's experience of spousal abuse, followed by the wife's age at cohabitation. [11] also observes that delaying a woman's age at first marriage reduces her likelihood of experiencing physical violence.

Table 3. Metric comparison across the models - Logistic regression, LASSO, RF, and XGBoost. Precision, Recall, F1-Score and AUC-PR are as defined in (1)

Model	Precision	Recall	F1-Score	AUC-PR
Logistic	0.735	0.710	0.722	0.789
LASSO	0.735	0.712	0.723	0.790
RF	0.870	0.801	0.834	0.797
XGBoost	0.872	0.775	0.820	0.789

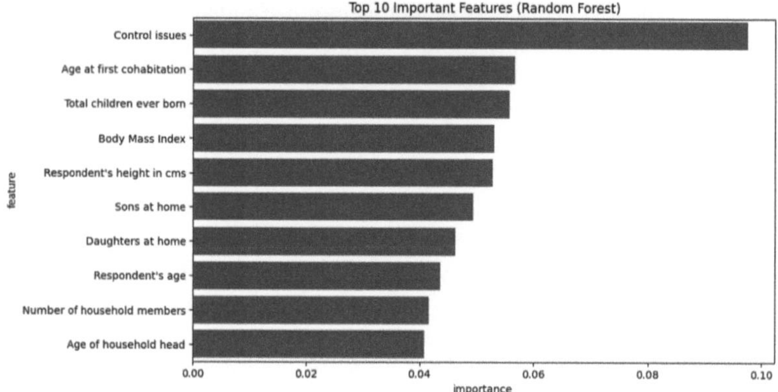

Fig. 2. Top 10 selected features based on importance ranking by Random Forest model

In addition, a higher number of children can lead to greater economic stress, which, in turn, increases the risk of spousal abuse. Women who are physically weaker, as captured by BMI and height, may also be more easily overpowered by their husbands [5]. Furthermore, a higher number of sons at home may provide greater bargaining power to women, as sons are more desirable than daughters in Indian society. As a result, a woman with more sons is less likely to experience spousal abuse.

While Random Forest identifies physical and demographic factors such as BMI, height, and age at cohabitation (directly linked to marriage) as key predictors, XGBoost highlights the additional relational and behavioral factors, such as intergenerational transmission of violence and mobile phone ownership. These insights suggest that effective interventions should target reducing controlling tendencies in partners, promoting education, and delaying early marriage. In addition, addressing intergenerational trauma and enhancing women's access to communication tools could help mitigate risk.

In addition, to understand the global and local interpretability of features and their direction, magnitude, and joint contributions to predictions, we analyze SHAP bar plots and summary plots. In the SHAP bar plot (Fig. 4), the features are ranked by their mean absolute SHAP values, indicating their overall contribution to the model's predictions across the entire dataset.

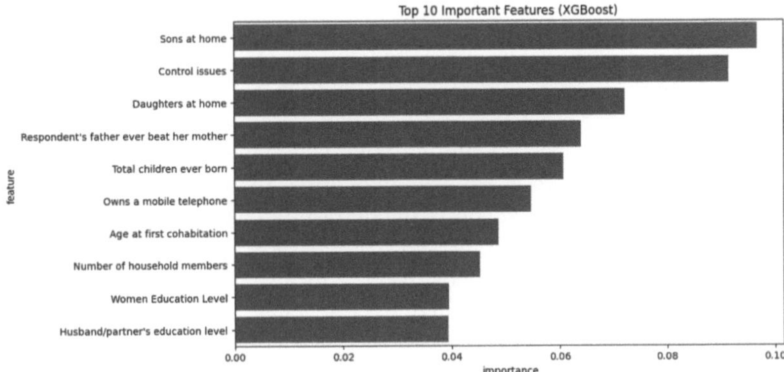

Fig. 3. Top 10 selected features based on importance ranking by XGBoost model

In line with the RF and XGBoost models, we find that the husband's control issue is the most important feature. Quite interestingly, from SHAP values, we find that a woman who justifies spousal abuse is more likely to experience abuse—a feature missing in the top 10 features given by the RF and XGBoost models. This may indicate that a woman's attitudes may not directly impact the prediction but may be interacting with other variables and captured by SHAP values. In addition, similar to XGBoost, we find strong support for the intergenerational transmission of abuse. Furthermore, as in RF and XGBoost predictions, we find that a woman's age at cohabitation is strongly related to her experience of spousal abuse. Finally, the husband's drinking habit strongly affects his acts of violence.

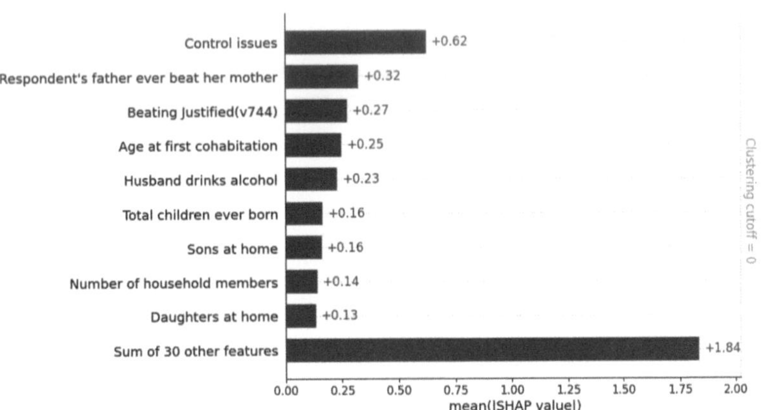

Fig. 4. SHAP bar plot ranking means absolute SHAP value of features, which quantifies their overall contribution to model predictions. The x-axis represents the average magnitude of SHAP, and the y-axis lists features in descending order of importance.

Furthermore, the SHAP summary plot, Fig. 5, shows the distribution of SHAP values for each feature. Each dot represents a single data point, and its position on the x-axis indicates the feature's impact on prediction (positive or negative SHAP value). Blue dots represent lower feature values, while red dots represent higher feature values. For the first feature, control issues of a husband, the high SHAP values (positive impact) are associated with higher predictions of the target class (violence). Justifying attitudes toward spousal abuse also shows a similar trend, where a woman's justification of violence pushes predictions toward a higher likelihood. Finally, the husband's drinking habit increases the risk of spousal abuse. In contrast, the woman's age at first cohabitation has a different distribution of blue and red dots. When the age at first cohabitation is lower, SHAP values tend to skew positively. This also suggests that younger ages at first cohabitation increase the likelihood of experiencing violence.

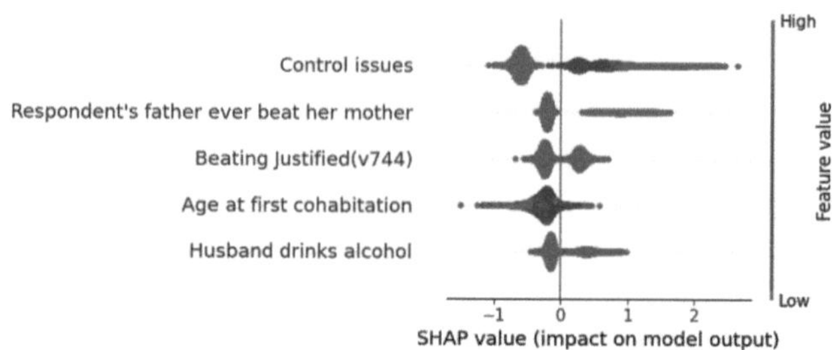

Fig. 5. SHAP summary plot. The y-axis lists features ranked by importance, and the x-axis represents SHAP values, indicating whether a feature increases or decreases the predicted likelihood of the target outcome. Each dot represents a data point, with color encoding showing the feature value (red = high, blue = low), helping to identify patterns such as whether higher or lower values of a feature contribute more to predicting domestic violence.

These national-level findings offer a foundational understanding of the drivers of domestic violence. However, given India's cultural diversity, the relative importance of these factors may vary regionally. In the following section, we explore these regional differences to identify context-specific patterns and interventions.

4.2   Regional Level Analysis

To understand the structural dynamics at various region levels, we repeat the same above analysis at each region level. The states covered in six geographic regions are outlined in Table 4.

First, we compare RF and XGBoost performance in Table 5 using metrics precision, recall, F1-Score, accuracy, and ROC-AUC as defined in Sect. 3.4. RF consistently achieves higher Recall, F1-Score, Accuracy, and ROC-AUC across all the regions compared to XGBoost. In terms of precision, XGBoost performed better in a few regions, but RF also performed comparably. Accuracy for RF was higher across regions, indicating better overall classification performance.

Next, in Table 6, we have the top 5 features for each of the six regions, as ranked by RF. We find that husbands with control issues are more likely to inflict violence on their wives. In addition, BMI emerges as the most common feature associated with domestic violence

Table 4. The classification of states and UTs across six regions of India, as given in the code–book of NFHS–5.

Region	States/UTs
North	Jammu and Kashmir, Himachal Pradesh, Punjab, Chandigarh, Uttarakhand, Haryana, NCT of Delhi, Rajasthan, Ladakh
Central	Uttar Pradesh, Chhattisgarh, Madhya Pradesh
East	Bihar, West Bengal, Jharkhand, Odisha
North-East	Sikkim, Arunachal Pradesh, Nagaland, Manipur, Mizoram, Tripura, Meghalaya, Assam
West	Gujarat, Dadra and Nagar Haveli and Daman and Diu, Maharashtra, Goa
South	Andhra Pradesh, Karnataka, Lakshadweep, Kerala, Tamil Nadu, Puducherry, Andaman and Nicobar Islands, Telangana

Table 5. Comparison of RF and XGBoost metrics across all regions. Precision, Recall, F1-Score, Accuracy, and ROC-AUC are as defined in (1)

	RF					XGB				
Region	Precision	Recall	F1-Score	Accuracy	ROC-AUC	Precision	Recall	F1-Score	Accuracy	ROC-AUC
North	0.87	**0.92**	**0.9**	**0.83**	**0.74**	**0.88**	0.88	0.88	0.8	0.69
Central	0.78	**0.84**	**0.81**	**0.74**	**0.7**	0.78	0.77	0.77	0.7	0.67
East	0.77	**0.85**	**0.81**	**0.74**	**0.71**	**0.78**	0.78	0.78	0.71	0.68
North-East	0.85	**0.89**	**0.87**	**0.8**	**0.73**	0.85	0.83	0.84	0.76	0.69
West	**0.85**	**0.89**	**0.87**	**0.8**	**0.74**	0.84	0.86	0.85	0.77	0.69
South	**0.78**	**0.78**	**0.78**	**0.73**	**0.72**	0.74	0.73	0.74	0.68	0.67

across regions. According to [5], a woman is more likely to experience violence if her husband can physically overpower her. Furthermore, a family history of intimate partner violence (IPV) is another significant factor, reflecting the intergenerational transmission of violence. This underscores the critical importance of addressing spousal abuse now to break the cycle of violence [17]. However, note that in South and West India, the age of the household head is the third most important feature, suggesting possible generational or authority-related influences in these areas. Women's age at first birth is another important factor.

To uncover within-region heterogeneity, we conduct clustering analysis. Figure 6 shows the heat map of clustering analysis using the K-means clustering algorithm on reduced data from the autoencoder. Each row represents a region, and each column represents a cluster based on the top 64 features selected by the autoencoder. The values in each row indicate the percentage of total data points from the region that fall into a particular cluster. The results reveal significant variation within regions.

For example, Central India has two dominant clusters: Cluster 1 (46%) and Cluster 3 (36%). This means that 82% of the Central India population lies within these two clusters. This suggests that relatively uniform policies can be implemented in this region.

Table 6. Top 5 features from RF across all the regions

	Feature 1	Feature 2	Feature 3	Feature 4	Feature 5
North	Control	BMI	Family history of IPV	Age of household head	Age at 1st birth
Central	Control	Family history of IPV	BMI	Age of household head	Age at 1st birth
East	Control	BMI	Family history of IPV	Age of household head	Respondent's age
North-East	Control	Family history of IPV	BMI	Age of household head	Age at 1st birth
West	Control	BMI	Family history of IPV	Age of household head	Husband drinks alcohol
South	Control	Family history of IPV	BMI	Age of household head	Age at 1st birth

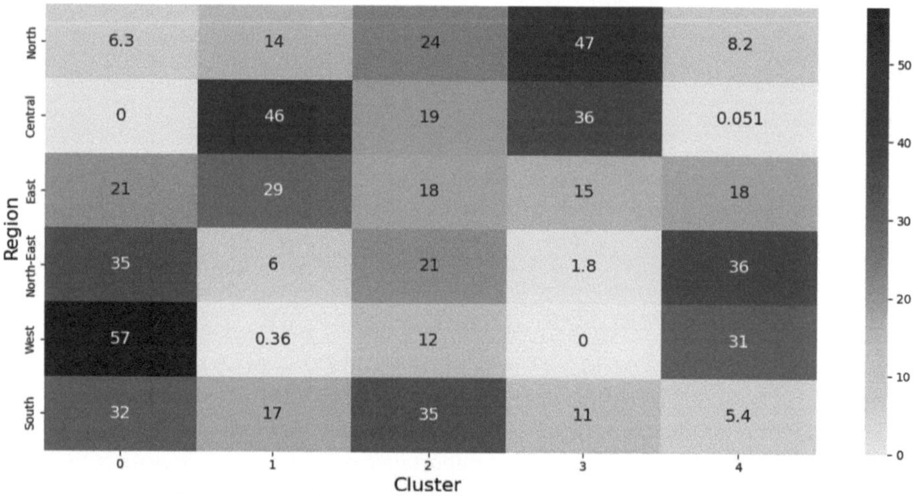

Fig. 6. Heatmap of clusters of data based on the top 64 features. Columns represent different clusters, and rows correspond to regions. The values in each row indicate the percentage of data points from that region assigned to each cluster, highlighting regional variations in feature distributions.

In addition, North India has a dominant presence in Cluster 3 (47%), followed by Cluster 2 (24%). Both North and Central India have the least presence in Cluster 0, highlighting a divergence from regions like South and West.

Meanwhile, in Cluster 0, South, West, Northeast, and East have a dominant presence. This implies some similarity between these regions. In contrast, for North and Central India, we find greater similarity between them, as a significant proportion of their populations lies in Cluster 3. East India appears diverse, with a relatively even distribution across multiple clusters. This suggests that tailored approaches are required to address the issue of domestic violence in this region.

5 Conclusion

In this paper, we examine the factors affecting Indian women's experience of spousal abuse, conditional on the region, using a nationally representative survey and machine learning techniques including, logistic regression, LASSO, RF, XGBoost, SHAP, and cluster analysis. RF demonstrates the finest balance of precision, recall, and F1-score, outperforming LASSO, Logistic, and XGBoost models in both the full sample and regional analyses.

The husband's control issues emerge as the most significant predictor across all analyses, reinforcing the need to address coercive control as a key determinant of spousal abuse. However, given the potential for unobserved confounding factors affecting both control issues and abuse, future research should consider employing instrumental variables or quasi-experimental designs to better isolate causal effects. We can also consider adding more variables to our analysis. In addition, we can check the areas within each of the six regions that are similar based on clustering analysis and suggest that even if a region is doing well overall, it may still have areas that can be improved.

In addition, findings highlight the intergenerational transmission of spousal abuse, emphasizing the need for interventions that break this cycle to prevent long-term harm to future generations. Clustering analysis further reveals significant intra-regional differences, with Central India showing relatively uniform patterns that may allow for standardized policy implementation, whereas East India exhibits more heterogeneous patterns, necessitating tailored, region-specific interventions.

Policy recommendations based on these findings include educational programs to address controlling behaviors, delaying cohabitation age through improved access to education, targeted interventions to break the intergenerational cycle of abuse, and economic empowerment programs to reduce financial stressors that contribute to domestic violence. In addition, region-specific policies should focus on vulnerable groups, such as women from families with a history of spousal abuse, while also addressing societal norms that justify domestic violence.

References

1. Rutstein, S.O., Rojas, G.: Guide to DHS Statistics. Calverton, MD: ORC Macro **38**, 78 (2006)
2. Bhattacharya, H.: Mass media exposure and attitude towards spousal violence in India. Soc. Sci. J. **53**(4), 398–416 (2016)
3. Sabia, J.J., Dills, A.K., DeSimone, J.: Sexual violence against women and labor market outcomes. Am. Econ. Rev. **103**(3), 274–278 (2013)
4. Dhar, D., et al.: Associations between intimate partner violence and reproductive and maternal health outcomes in Bihar, India: a cross-sectional study. Reprod. Health. Health **15**, 1–14 (2018)
5. Eswaran, M., Malhotra, N.: Domestic violence and women's autonomy in developing countries: theory and evidence. Canadian Journal of Economics/Revue canadienne d''economique **44**(4), 1222–1263 (2011)

6. United Nations: Ending violence against women and girls: Programming essentials. Retrieved September, 2, p. 10 (2013)
7. Dalal, K., Lindqvist, K.: A national study of the prevalence and correlates of domestic violence among women in India. Asia Pacific J. Public Health **24**(2), 265–277 (2012)
8. Pakrashi, D., Saha, S.: Intergenerational consequences of maternal domestic violence: Effect on nutritional status of children. GLO Discussion Paper (2020)
9. Jejeebhoy, S.J.: Associations between wife-beating and fetal and infant death: impressions from a survey in rural India. Studies in Family Planning, 300–308 (1998)
10. Begum, S., Donta, B., Nair, S., Prakasam, C.P.: Sociodemographic factors associated with domestic violence in urban slums, Mumbai, Maharashtra, India. Indian J. Med. Res. **141**(6), 783 (2015)
11. Roychowdhury, P., Dhamija, G.: The causal impact of women's age at marriage on domestic violence in India. Feminist Econ. **27**(3), 188–220 (2021)
12. McDougal, L., Dehingia, N., Bhan, N., Singh, A., McAuley, J., Raj, A.: Opening closed doors: using machine learning to explore factors associated with marital sexual violence in a cross-sectional study from India. BMJ Open **11**(12), e053603 (2021)
13. Tauchen, H.V., Witte, A.D., Long, S.K.: Violence in the family: a non-random affair. Int. Econ. Rev. **32**(2), 491–511 (1991)
14. Anderberg, D., Mantovan, N., Sauer, R.M.: The dynamics of domestic violence: learning about the match. Econ. J. **133**(656), 2863–2898 (2023)
15. Dhanaraj, S., Mahambare, V.: Male backlash and female guilt: women's employment and intimate partner violence in urban India. Feminist Econ. **28**(1), 170–198 (2022)
16. Palermo, T., Bleck, J., Peterman, A.: Tip of the iceberg: reporting and gender-based violence in developing countries. Am. J. Epidemiol.Epidemiol. **179**(5), 602–612 (2014)
17. Sharma, M., Stern, S.: Simultaneous Hazard Rate Estimation of First Incident of Spousal Abuse and First Birth. Unpublished Manuscript (2024)
18. Sharma, M., Stern, S.: Generalized Weibull Distributions. Working Paper No. 24-05, Stony Brook University, Department of Economics (2024)
19. Luca, D.L., Owens, E., Sharma, G.: Can alcohol prohibition reduce violence against women? Am. Econ. Rev. **105**(5), 625–629 (2015)
20. Shashidhara, S., Mamidi, P., Vaidya, S., Daral, I.: Using machine learning prediction to create a 15-question IPV measurement tool. J. Interpersonal Violence **39**(1–2), 11–34 (2024)
21. Sardinha, L., Maheu-Giroux, M., St¨ockl, H., Meyer, S.R., Garc´ıa-Moreno, C.: Global, regional, and national prevalence estimates of physical or sexual, or both, intimate partner violence against women in 2018. The Lancet **399**(10327), 803–813 (2022)
22. Jejeebhoy, S.J.: Wife-beating in rural India: a husband's right? Evidence from survey data. Economic and Political Weekly, 855–862 (1998)
23. Merlo, J., Chaix, B., Yang, M., Lynch, J., R°astam, L.: A brief conceptual tutorial of multilevel analysis in social epidemiology: linking the statistical concept of clustering to the idea of contextual phenomenon. J. Epidemiol. Community HealthEpidemiol. Community Health **59**(6), 443–449 (2005)
24. Lenze, J., Klasen, S.: Does women's labor force participation reduce domestic violence? Evidence from Jordan. Feminist Economics **23**(1), 1–29 (2017)
25. World Health Organization: Violence against women prevalence estimates, 2018: global, regional, and national prevalence estimates for intimate partner violence against women and global and regional prevalence estimates for non-partner sexual violence against women. Executive summary. Geneva: World Health Organization (2021)
26. Mahapatro, M., Gupta, R.N., Gupta, V.: The risk factor of domestic violence in India. Indian J. Commun. Med. **37**(3), 153–157 (2012)
27. Bloch, F., Rao, V.: Terror as a bargaining instrument: a case study of dowry violence in rural India. Am. Econ. Rev. **92**(4), 1029–1043 (2002)

28. Bobonis, G.J., Gonz´alez-Brenes, M., Castro, R.: Public transfers and domestic violence: The roles of private information and spousal control. American Economic Journal: Economic Policy **5**(1), 179–205 (2013)

29. Dutta, N., Rishi, M., Roy, S., Umashankar, V.: Risk factors for domestic violence—an empirical analysis for Indian states. J. Dev. Areas, 241–259 (2016)

30. Koenig, M.A., Stephenson, R., Ahmed, S., Jejeebhoy, S.J., Campbell, J.: Individual and contextual determinants of domestic violence in North India. Am. J. Public Health **96**(1), 132–138 (2006)

31. Chatterjee, S., Poddar, P.: Women's empowerment and intimate partner violence: evidence from a multidimensional policy in India. Econ. Dev. Cultural Change **72**(2), 801–832 (2024)

32. Anderberg, D., Rainer, H., Wadsworth, J., Wilson, T.: Unemployment and domestic violence: theory and evidence. Econ. J. **126**(597), 1947–1979 (2016)

33. Aizer, A.: The gender wage gap and domestic violence. Am. Econ. Rev. **100**(4), 1847–1859 (2010)

34. Panda, P., Agarwal, B.: Marital violence, human development and women's property status in India. World Dev. **33**(5), 823–850 (2005)

35. Farmer, A., Tiefenthaler, J.: An economic analysis of domestic violence. Rev. Soc. Econ. **55**(3), 337–358 (1997)

36. Anderberg, D., Rainer, H.: Economic abuse: a theory of intrahousehold sabotage. J. Public Econ. **97**, 282–295 (2013)

37. Brahma, D., Mukherjee, D.: Early warning signs: targeting neonatal and infant mortality using machine learning. Appl. Econ. **54**(1), 57–74 (2022)

38. Ram, A., Victor, C.P., Christy, H., Hembrom, S., Cherian, A.G., Mohan, V.R.: Domestic violence and its determinants among 15–49-year old women in a rural block in South India. Indian J. Community Med. **44**(4), 362–367 (2019)

39. Kaur, R., Garg, S.: Addressing domestic violence against women: an unfinished agenda. Indian J. Community Med. **33**(2), 73 (2008)

40. Tambe, M., Rice, E.: Artificial Intelligence and Social Work. Cambridge University Press (2018)

41. Amusa, L.B., Bengesai, A.V., Khan, H.T.A.: Predicting the vulnerability of women to intimate partner violence in South Africa: Evidence from tree-based machine learning techniques. J. Interpersonal Violence **37**(7–8), NP5228–NP5245 (2022)

42. Jayachandran, S.: Social norms as a barrier to women's employment in developing countries. IMF Econ. Rev. **69**(3), 576–595 (2021)

43. Lundberg, S.M., Lee, S.-I.: A unified approach to interpreting model predictions. In: Guyon, I., Luxburg, U.V., Bengio, S., Wallach, H., Fergus, R., Vishwanathan, S., Garnett, R.: (eds.) Advances in Neural Information Processing Systems 30, pp. 4765–4774 (2017)

Author Index

Y. Folajimi et al. (Eds.): SIAI 2025, CCIS 2599, p. 143, 2025.
https://doi.org/10.1007/978-3-031-98949-0

MIX
Papier aus verantwortungsvollen Quellen
Paper from responsible sources
FSC® C105338

If you have any concerns about our products,
you can contact us on
ProductSafety@springernature.com

In case Publisher is established outside the EU,
the EU authorized representative is:
**Springer Nature Customer Service Center GmbH
Europaplatz 3, 69115 Heidelberg, Germany**

Printed by Libri Plureos GmbH
in Hamburg, Germany